UNWIN HYMAN COLLECTIONS

NORTHERN LIGHTS

INCLUDING
FOLLOW ON
ACTIVITIES

EDITED BY LESLIE WHEELER

AND DOUGLAS YOUNG

Published in 1989 by
Unwin Hyman Limited
15–17 Broadwick Street
London W1V 1FP

Typeset by TJB Photosetting Ltd., Grantham, Lincolnshire
Printed in Great Britain by Billings & Sons Ltd., Worcester

Series Editor: Roy Blatchford
Series cover design by Iain Lanyon
Cover illustration by Elizabeth Haines

CONTENTS

INTRODUCTION

Standard Grade English, after a long gestation period and a birth that was far from easy, is now alive and well, and finding its feet in classrooms throughout Scotland.

Although the principles that lie behind this development are rooted very firmly in what has been good practice for many years, there are changes of emphases, and new responsibilities placed upon teachers to design their own courses in response to the particular needs of their pupils. It is hoped that this book will help teachers to cope with these demands.

The centrality of literature to any English course is given new emphases; as is the need for this literature to contribute to the personal enrichment of the pupils by reflecting the cultural diversity of life in Scotland today. This collection responds to these emphases by providing some examples of very recent Scottish writing which explore the complexity of contemporary Scottish life, particularly for young people.

Literature is also important in providing contexts for writing, and particularly for talking. The recognition of the importance of talk in the English classroom is perhaps the most important of the new emphases, not only because of its own inherent value, but also because talking about a text is seen as a key way in which pupils can be actively involved in constructing meaning rather than being passive receivers of the teacher's reading of the text. We have tried to pay particular heed to these connections between reading, writing and talk in our suggestions for follow on activities.

It has become a feature of this series to include personal essays by some of the writers, in which they explain how they came to write their stories. The insight which this gives pupils into the processes of writing has led us to continue this practice.

Finally, whilst we have been mainly concerned with providing ready material for Scottish children at a certain stage in their school careers, we do believe that each of the pieces is an example of good literature and as such will appeal to readers of all ages, Scottish or non-Scottish.

Douglas Young
Leslie Wheeler

CONVERSIONS

Old Wives' Tales

That, kiddo, was called a contraction. No doubt about it now.

So. This is it, then.

It's just you and me now. Till birth us do part.

Don't get scared. I reckon we have a few hours left to ourselves before we have to tell anyone. And I need the time. Okay? You see, while the world is turning ever so casually on its axis, I am about to be slipped into a supporting role for the rest of my life to leave you the centre stage. It may be the first day of your life, but it's the last day of my childhood. So give me a few hours to get used to the idea, eh?

That sounds selfish. I know. But don't fret. There'll be years and years ahead for me to worry about you.

I must say you haven't got much of a day for it. In fact, if I were at all superstitious and chose to believe the omens like grandma did, I can tell you they're not good. The sun's not even shining. There's a distinct smell of withering in the air this grey September morning. See, the roses are rotted with the rain and it looks like every pest in the book is busy chewing holes and spitting green. At least Fat-Cat's pleased. They've cut the barley and all the refugee mice have taken him completely by surprise this morning. He'll come in and present us with another little furry body in some advanced state of shock any time now. Stay where you are for a while. Have a last swim around in the warm dark. There's nothing out here for you today. No star in the east or three wise men for you, kiddo.

Funny how I thought I'd be ready for all this. Just a few more hours. Please? If you don't panic, I won't. But it's not how it was described in the manual.

Aww...Jesus! Okay, okay. I'll see if I can get you one wise woman.

I told you. The world out here's not a place to rush into.

There's AIDS, drugs and God knows what all lurking out here, colds and rotten teeth, acid rain and nuclear waste — all this in spite of the wonder of the modern world. (They can predict the return of Halley's comet with pinpoint accuracy, but they couldn't tell me when you would be born. Doesn't that seem odd to you?)

What else can I tell you about your chosen birth day, then? Well, Nelson Mandela is *still* in prison, there are killings in Northern Ireland, wars in three continents, famine in another, but, Coronation Street carries on. Women are spying on Greenham Common and picking brambles at the same time, pinning hopes on wholesome pies and laying down a future in jams and good strong wine. Just wait till you taste a bramble and apple pie!

It's a funny old world. And don't you go expecting me to explain it all away, either. I'll wipe your bum and put plasters on your knees, maybe teach you to play the piano and help you with your homework. After that, you're on your own, kiddo. Like the rest of us. Mummies can't kiss the world better any more and there ain't no fairy godmothers.

There are grandmas, though. They're the next best thing. They're good with the fairy tales, are grandmas. I have one of my own, so I know. Of course, you'll be seeing her soon. She's something else, that one. When I asked her — you know, woman to woman for the very first time — what it was like the day mother was born, she told me it was snowing! The omens seemed good for the times, she said. It was Winston Churchill's birthday and St Andrew's Day, so the flags were flying. Well, you had to keep your spirits up during the war, she said. She would tell my mother the flags were up for her and not worry her about things like war.

You could keep them ignorant then, you see. No TV. Anyway, she told me all of that, almost in one breath, but did not volunteer one word about the birth itself. She remembers the midwife's jokes and her bad feet, though. She patted my arm finally and said I'd forget about it as soon as it was over. That was a great help. And mother, she still can't bear to watch war documentaries or read about the Holocaust because she knows she spent the whole sorry six years skipping through other people's time. I teased her about grandma telling her the flags were for her and she gave me that look. The one that can tidy up a room.

Not that she was any great help either. When I asked her what it was like the day I was born, she said it was a scorcher and she just about died of thirst. My God! I should have thought that was the least of her problems. I persisted, but it's not easy asking you mother little questions like — how was your labour? She told me it took twenty-eight hours and the temperature outside rose to eighty at one point. Some kids at the local school sports collapsed with heatstroke, she remembers. I wasn't asking about the weather, dammit.

If you're a girl, I promise, I do, I promise that when the time comes I'll tell you every last detail. Come to think of it, I'll tell you even if you're a boy. I'm all for equality.

In the end, she confided it wasn't too bad. Anyway, she said, it'll all be worth it. And that's it. That's the extent of the experience handed down through generations to me, now, with my big belly.

Hey! Is this hurting you as much as it's hurting me? Read the book, kiddo. This is supposed to be the good bit. So hang on to your hat, because here it comes again.

Holy. Hell!

I knew it. I just knew it. She was lying through her teeth. Twenty-eight hours she was in labour with me. Twenty-eight hours! And all she'll tell me is that it wasn't too bad. So what's bad? What are we registering on the Richter scale now, for example? I tell you, this is getting scary.

Where's that book?

Where's that bloody book?

Recite something, the book says. Quick.

> Humpty Dumpty sat on a wall
> Humpty Dumpty had a great fall
> All the King's horses
> And all the King's men
> Couldn't put Humpty together again.

That's going to go down well in the ambulance. Speaking of which, I'm going to phone them. Sorry, but I'm losing my grip, as they say, and it's your fault.

No, forgive me. That's not true. It's their fault. The secret society of mothers out there.

You okay in there? Just hold on a while yet, kiddo. Hey! I'll soon have to stop calling you that. You'll have a name. And all the other trappings of a statistic. They'll start a file on you and you'll get orange juice, an education and free dental treatment till you're sixteen. Then you'll get to collect Social Security like the rest of us. Just like your Dad. And if you're wondering where your other parent is at this critical juncture in your life, he's out looking for work, again. That's where he is. I sent him off this morning without telling him about the queer ache in my back that started all this. Well, he would only have fussed. He's a lovely man, though. You'll like him. I promise. But, if he doesn't find a job soon, you'll have that on your birth certificate, you know. Father — Unemployed.

This is not funny any more. I'm not sure I can take much more of it. WHY DIDN'T SOMEONE TELL ME?

> Humpty Dumpty sat on a wall
> Humpty Dumpty had a great fall…

Myself, I think he was conned. Just like me. Why else would an egg go and jump off a wall? Well, there's a lot of it about. I suppose. Conspiracy, that is. I did warn you it wasn't much of a world. Everybody's at it. Governments, drug companies, slimming magazines and that silly bitch who ran the antenatal classes. To think I believed all that guff about clenched fists mimicking contractions.

Remember? 'Tighten. Clench. Ho-o-ld it. Re-lax.' What a load of crap.

Just like mother and her 'not too bad'. Just like all of them, all the bloody mothers down all the bloody generations who have never told it as it is. THANK YOU FOR NOTHING ALL YOU MOTHERS OUT THERE.

Oh, but it will all be worth it, they say.

Oh? Will it really?

You'll forget about it as soon as it's over, they say.

No I won't. I'm going to spill the beans. I'm going to tell all. And I'm telling you, kiddo. This is hell.

This. This is the biggest con of them all. This is the lulu. This...Oh, God...I bet if someone asked the Virgin Mary what it was like the day Jesus was born, she'd say it was a fine clear night with plenty of stars.

I wonder if Jesus had a birth certificate? Father — God. Hah! Imagine trying to get that through the system today.

INPUT ERROR.

PLEASE RETYPE.

Don't worry, kiddo. You don't have a God in the family. You won't upset any of their computer programmes.

> Humpty Dumpty sat on a wall
> Humpty Dumpty had a great fall
> All the King's horses and all the King's men
> Couldn't put Humpty together again.

It has not escaped my notice how much I resemble an egg.

So.

It's time to go, kiddo.

Let's go jump off the wall.

Essay

'Old Wives' Tales' is the story of a woman about to give birth. It is written from her viewpoint and is, in fact, a monologue, since the person she is speaking to is her unborn child. The woman is caught in a moment of time just before her life is about to be changed forever. She is using that time to commentate on the world her baby will be born into. The world does not appear to be a very welcoming place for a new life, at least not the way this mother views it, yet the tone of the story is not intended to make the reader sad or depressed. In spite of everything, it is meant to leave a note of hope in the reader's mind, because the woman is strong — a survivor who can take the world situation in her intellectual stride, even if, as the story suggests, the surprise her body has in store for her might be that bit more difficult to handle.

This story is not autobiographical, except that it echoes my own surprise at the intensity of the birth experience, in spite of all the books I'd read on the subject. (I was a classic case of thinking I knew all I needed to know, because I had read 'the book'.) And I could not have written the story at the time my children were born, because the experience would have been too close and too intimate to try to analyse. Even years after, when I was writing the story, I found it difficult to control the tone and stop it drifting into deep, dramatic trouble. Birth as a subject does not lend itself very easily to the light and humorous touch, which is exactly why I chose to write the story that way, to make it different.

A few ideas came together to set the story going in my head. A young friend who was going to have a baby asked me what it was going to be like and I found myself confused about what and what not to tell her. I ended up spouting the same old platitudes my own mother had used with me, even while I knew they would be just as little use. That was when I began thinking that motherhood must be both the largest and the most exclusive club in the world, with an initiation ceremony — childbirth — still wrapped in enough mystery so that even well-informed twentieth-century women can be taken by surprise, because mothers themselves, for whatever reason, have chosen to keep quiet about it.

There are plenty of factual books written about birth, and hundreds of fictional story-lines which hinge on the drama of that major life event. However, these books could be said to share a common view — the view from around the bed, the spectators' view — not the view from the bed itself. It is far less easy to find fictional accounts of childbirth written from the mother's viewpoint. I began to wonder about that and decided to try to write about the experience from the inside, as it were. I also wanted to try to change the 'go and boil up lots of hot water' approach to birth which is still found in some books and films. I used to wonder what they needed all that water for, anyway. I knew a baby's bath was only a couple of inches deep! Of course, I know now that it's for washing the bed-sheets afterwards, but that's another thing I had to find out for myself.

I find it interesting, with hindsight, that this story and another of mine called 'Five Days of Clean Sheets' (dealing with that time in hospital after a baby is born), which is also written from the mother's viewpoint, could be put together and what would be missing would be an account of an actual birth. I would still find that quite impossible to write about, all these years later. Which is to say that perhaps this goes a little way towards explaining why so few first-hand accounts of birth are found in even modern fiction. Like me, perhaps, most women would prefer to leave writing about the birth process to the most gifted of women writers with the experience to draw on.

The monologue style of this story allowed me to bring the mother's thoughts out into the open without having them related by an unseen narrator, which would not have been appropriate for so intimate an experience. It also solved the problem of which tense to use to tell the story. The mother is speaking throughout, even if there are no quotation marks, and that gives the story the feeling of happening now while still allowing the mother to reflect on the past. As a writer this gives me the best of both worlds and a freedom not to be found to the same extent in a story with an unseen narrator. I enjoy using this style now and again and find I turn to it when I really need to get right inside a character's head.

If you are interested in trying to write short stories, the monologue approach is only one of a dozen possible approaches, but it is perhaps one of the simplest to start on. So, too, is choosing to write about a person's feelings before or after a dramatic event instead of the more usual account of the event itself, to make your story different from what others on the same topic are likely to be about.

Short stories are the only form of fiction I write, but I feel I could go on exploring the short story form for years without exhausting the possible approaches. As a form of writing I find it fascinating, but I discovered early on that short stories are not easy to write because they are short. They are *difficult* to write because they are short.

The Doctor

Yes, I agreed to perform the abortion.
The girl was under unusual strain.
I formed the opinion that for personal reasons
and home circumstances her health would suffer
if pregnancy was not terminated.
She was unmarried and the father was unknown.
She had important exams to sit,
her career would be jeopardized, and in any case
she went in mortal fear of her father
(who is himself, as it happens, a doctor)
and believed he would throw her out of the house.
These factors left me in no doubt.
Accordingly I delivered her seven months baby
without complications. It was limp and motionless.
I was satisfied there was no life in it.
Normal practice was followed: it was placed
in a paper disposal bag and sent
to the incinerator. Later to my surprise
I was told it was alive. It was then returned
and I massaged its chest and kept it warm.
It moved and breathed about eight hours.
Could it have lived? I hardly think so.

You call it a disturbing case? Disturbing
is a more emotive word than I would choose
but I take the point. However, the child
as far as I was concerned was dead
on delivery, and my disposal instructions
were straight and without melodrama.
There is, as sheriff and jury will agree,
an irony for students of the human condition
(and in this case who is not?)
in the fact that the baby was resuscitated
by the jogging of the bag on its way to the incinerator.
I hope that everything I have said is clear.

PAT GERBER

Shadows Fall

The sound of the kettle strains into the silence. A floorboard creaks under my pink-slippered foot. I go through the comforting routine of tea-brewing; real leaves,none of your old bags for me. Especially not on a morning like this, after a night like that.

In the dark hall behind me chimes the grandmother clock, brassily announcing 5 a.m. And he's still up there. What ought I to do? Something, surely, if I'm a mother worth the name.

All night I'd lain awake thinking of Marie my daughter. I'd tried to visualise what my mother would have done in similar circumstances. I failed. The situation simply could not have arisen.

Things had been different then. And in some ways worse. There had been a curfew operating in our house when I was seventeen. Home at nine most nights. My young men had seen me home, given me an inexpert peck in a doorway long before we got there, and left the minute my father opened the front door. I had learned, painfully, that my father had to be obeyed. With apparent docility I evolved a way of surviving his rules. So strong was the call of young life being lived out there in the real world, I often listened for his snores, then shinned down a well-placed drainpipe to rejoin whatever party was going on. Johnny and I had slept rough under the trees. Then I'd clamber back up the pipe in the small hours, certain that my father would hear my very heartbeats. He never did. In the end poor Johnny had to marry me. It was the way in those days, when a girl fell pregnant.

Now, in these new small hours, is Marie my daughter making the same mistake? My ears stretch into the dark house. But there are no — sounds.

I look again at the note. She'd left it on the kitchen table for me and I'd found it when I'd come in from the reunion with

10

school-friends of yesteryear. It was beautifully written in her round neat hand, as usual.

'Hi Mum,' it said. 'I'm in. Hope you had a good time with the Ancients. 2 Alkaseltzer below attached. Mike's staying the night in my room. I'll be leaving early for swimming practice — try not to waken you. Night, sleep well. XXX.'

'Now now, young man, you must get up and leave this house immediately. Never darken our door again.' Should I have marched into her room and delivered such a speech?

Or, at seventeen, was she now an adult, free to make her own choices, her own mistakes?

She had been my mistake. Not my only one, but the one with the farthest-reaching consequences.

I'd had a struggle to rear her after Johnny left home. We'd managed, with difficulty; I was always faced with the choice of going out to work and leaving her as a child alone in the house if she was ill or on holiday, or of staying at home and going without the low part-time wage that fed and clothed us and paid the rent. She'd been a difficult, resentful creature in her middle teens. But now she was an assured, hard-working, frank young woman.

So frank that I knew her views on religion — no such thing as God; politics — she was the last feather on the edge of the left-wing; and sex, no pill yet, but she'd start taking it when she thought it was time.

I, knotted with inhibition, gnarled with guilt on the subject of anything physical, had struggled to inform her on the bare facts of life. A book, tastefully coloured pink-for-girls, left lying around, had, I hoped, filled in the gaps. I hadn't been able to manage words like masturbation or penis.

Drink the tea. Calm streaking nerves. Turn on the radio softly to blank out thought. That's it, gentle music. Six chimes on the grandmother clock. He's still up there, that spiky-headed boy. In Marie's — bed? It would be utterly naive to imagine in these permissive days that he'd be anywhere else, wouldn't it? I must steel myself for the worst.

Marie had introduced him to me last December before the school Christmas party. Now he was in sixth year with her. He was just one of her large gang of friends. They'd usually spend

Friday or Saturday evenings together in one or other of their family homes, playing music, dancing a little, talking, drinking lager. Often, when it was our turn, one or two of her girlfriends would stay the night; she'd do the same after an evening at theirs. Marie'd never mentioned this boy, Mike, particularly — or had I just not been listening?

Sometimes, as she rattled through all the minutiae of her day, or the night before's party, my mind would slide away to dwell on some problem of my own — a bill I couldn't pay, a lump I didn't want to notice in my breast.

Marie had her own life to live. With a bit of luck she'd make a better job of hers than I had of mine. Things were easier now, for a girl. She'd do well in her exams, perhaps get into University to study Pharmacy. Her Saturday job in the chemist would help her eke out her grant. She'd have a really good career ahead of her, there would be no need to hunt for some unreliable man to keep her. No need to hurry into marriage. No need to marry at all? And children?

She was frank about children too. Couldn't abide them. Said she for one wasn't going to add to the overpopulation of the world and have her life eaten up by any puking, leaky brat. But I remember how tenderly she loved her hard plastic, unresponsive doll. She had a warm well of love to give, to someone, some day. Even though it was protected these days by the thorns and pins of fashionable punkery.

Now I must find courage to go, must go treading up these stairs, must go into her bedroom and stop, yes stop this spikey Michael from doing any more, any further harm to my own, my darling daughter.

Strong anger lights my way. Pink suede hushes my tread on the stairs. Outside Marie's door I listen and feel like a criminal.

Alarm bells scream. I back away, turn and run, weeping hot tears down again to the kitchen. I wait, heart thumping.

Someone is coming, is coming downstairs. How will I face him, them? There is a long yawn.

'Any tea left in the pot Mum? What're you doing up so early?'

She's wearing the worn-out pink kaftan she 'borrowed' from me ages ago.

'Your note,' I mumble. 'Michael?'

'Oh he'll not want any tea. He's out for the count. Hope you didn't mind — bunged him in the sleeping bag, he's so thick he'll not feel the floor! I'll take him a coffee when I get back from the baths, OK? Yuk this tea's cold — I'll make a fresh pot. You look awful. Must've been quite a party! The Ancients all still surviving? Oh, Jenny and I are going to watch the rugger this morning at ten — Mike's playing for the first, and Andy and Jim. Then we're all going back to Jenny's tonight....'

The kettle whistles. Outside the sun is beginning to rise. A blackbird sings on the clothes pole.

Poem for my Sister

My little sister likes to try my shoes,
to strut in them,
admire her spindle-thin twelve-year-old legs
in this season's styles.
She says they fit her perfectly,
but wobbles
on their high heels, they're
hard to balance.

I like to watch my little sister
playing hopscotch, admire the neat hops-and-skips of her,
their quick peck,
never-missing their mark, not
over-stepping the line.
She is competent at peever.

I try to warn my little sister
about unsuitable shoes,
point out my own distorted feet, the callouses,
odd patches of hard skin.
I should not like to see her
in my shoes.
I wish she could stay
sure footed,
          sensibly shod.

F *light*

One day off-handedly her sister told her
she was fat. Sixteen years old,
construing her adolescence in the mirror
and deciding she objected to the mould
and curve of breast and thigh,
she took to muesli, oranges and lettuce,
lost thirty pounds in sixty days,
tucking in jeans instead of food. 'Don't fuss',
she told her mother, who called it 'Just a phase'.
Father, table-thumping, predicted she would die.

Thin as a sparrow, secretive as a water-rail,
with pointed nose and elongated toes,
she pecked her food, grew pale,
sprouted fingered wings and rose
one day from family noise and bother
to flutter through the open kitchen window,
circling and soaring high into the sky.
Father preoccupied failed to see her go.
Mother remarked, 'I'm not surprised. Our Di
ate just like a bird.' 'How absurd!' said Father.

Mother it was who missed her most, pattering
round the house beslippered every evening,
leaving the cage-door open and scattering
bird-seed on a plate. But too little and too late.

The Lost Boy

There was one light in the village on Christmas Eve; it came from Jock Scabra's cottage, and he was the awkwardest old man that had ever lived in our village or in the island, or in the whole of Orkney.

I was feeling very wretched and very ill-natured myself that evening. My Aunty Belle had just been explaining to me after tea that Santa Claus, if he did exist, was a spirit that moved people's hearts to generosity and goodwill; no more or less.

Gone was my fat apple-cheeked red-coated friend of the past ten winters. Scattered were the reindeer, broken the sledge that had beaten such a marvellous path through the constellations and the Merry Dancers, while all the children of Orkney slept. Those merry perilous descents down the lum, Yule eve by Yule eve, with the sack of toys and books, games and chocolate boxes, had never really taken place at all.... I looked over towards our hearth, after my aunt had finished speaking: the magic had left it, it was only a place of peat flames and peat smoke.

I can't tell you how angry I was, the more I thought about it. How deceitful, how cruel, grown-ups were! They had exiled my dear old friend, Santa Claus, to eternal oblivion. The gifts I would find in my stocking next morning would have issued from Aunty Belle's 'spirit of generosity'. It was not the same thing at all. (Most of the year I saw little enough of that spirit of generosity — at Halloween, for example, she had boxed my ears till I saw stars that had never been in the sky, for stealing a few apples and nuts out of the cupboard, before 'dooking' time.)

If there was a more ill-tempered person than my Aunty Belle in the village, it was, as I said, old Jock Scabra, the fisherman with a silver ring in his ear and a fierce one-eyed tom cat.

His house, alone in the village, was lit that night. I saw it, from our front door, at eleven o'clock.

Aunty Belle's piece of common sense had so angered me, that I was in a state of rebellion and recklessness. No, I would *not* sleep. I would not even stay in a house from which Santa had been banished. I felt utterly betrayed and bereaved.

When, about half past ten, I heard rending snores coming from Aunty Belle's bedroom, I got out of bed stealthily and put my cold clothes on, and unlatched the front door and went outside. The whole house had betrayed me — well, I intended to be out of the treacherous house when the magic hour of midnight struck.

The road through the village was deep in snow, dark except where under old Scabra's window the lamplight had stained it an orange colour. The snow shadows were blue under his walls. The stars were like sharp nails. Even though I had wrapped my scarf twice round my neck, I shivered in the bitter night.

Where could I go? The light in the old villain's window was entrancing — I fluttered towards it like a moth. How would such a sour old creature be celebrating Christmas Eve? Thinking black thoughts, beside his embers, stroking his wicked one-eyed cat.

The snow crashed like thin fragile glass under my feet.

I stood at last outside the fisherman's window. I looked in.

What I saw was astonishing beyond ghosts or trows.

There was no crotchety old man inside, no one-eyed cat, no ingrained filth and hung cobwebs. The paraffin lamp threw a circle of soft light, and all that was gathered inside that radiance was clean and pristine: the cups and plates on the dresser, the clock and ship-in-the-bottle and tea-caddies on the mantelpiece, the framed picture of Queen Victoria on the wall, the blue stones of the floor, the wood and straw of the fireside chair, the patchwork quilt on the bed.

A boy I had never seen before was sitting at the table. He might have been about my own age, and his head was a mass of bronze ringlets. On the table in front of him were an apple, an orange, a little sailing ship crudely cut from wood, with linen sails, probably cut from an old shirt. The boy — whoever he was — considered those objects with the utmost gravity. Once he put out his finger and touched the hull of the toy ship; as if it was so precious it had to be treated with special delicacy, lest it broke like a soap-bubble. I couldn't see the boy's face — only his bright hair, his lissom neck, and the gravity and joy that informed all his gestures. These were his meagre Christmas presents; silently he rejoiced in them.

Beyond the circle of lamp-light, were there other dwellers in the house? There may have been hidden breath in the darkened box bed in the corner.

I don't know how long I stood in the bitter night outside. My hands were trembling. I looked down at them — they were blue with cold.

Then suddenly, for a second, the boy inside the house turned his face to the window. Perhaps he had heard the tiny splinterings of snow under my boots, or my quickened heart-beats.

The face that looked at me was Jock Scabra's, but Jock Scabra's from far back at the pure source of his life, sixty winters ago, before the ring was in his ear and before bad temper and perversity had grained black lines and furrows into his face. It was as if a cloth had been taken to a tarnished web-clogged mirror.

The boy turned back, smiling, to his Christmas hoard.

I turned and went home. I lifted the latch quietly, not to awaken Aunty Belle — for if she knew what I had been up to that midnight, there would have been little of her 'spirit of generosity' for me. I crept, trembling, into bed.

When I woke up on Christmas morning, the 'spirit of the season' had loaded my stocking and the chair beside the bed with boxes of sweets, a Guinness Book of Records, a digital watch, a game of space wars, a cowboy hat, and a 50 pence piece. Aunty Belle stood at my bedroom door, smiling. And, 'A merry Christmas,' she said.

Breakfast over, I couldn't wait to get back to the Scabra house. The village was taken over by children with apples, snowballs, laughter as bright as bells.

I peered in at the window. All was as it had been. The piratical old man sluiced the last of his breakfast tea down his throat from a cracked saucer. He fell to picking his black-and-yellow teeth with a kipper-bone. His house was like a midden.

The one-eyed cat yawned wickedly beside the new flames in the hearth.

JAMES KELMAN

Away in Airdrie

During the early hours of the morning the boy was awakened by wheezing, spluttering noises and the smell of a cigarette burning. The blankets hoisted up and the body rolled under, knocking him over onto his brother. And the feet were freezing, an icy draught seemed to come from them. Each time he woke from then on he could either smell the cigarette or see the sulphur head of the match flaring in the dark. When he opened his eyes for the final time the man was sitting up in bed and coughing out: Morning Danny boy, how's it going?

I knew it was you.

Aye, my feet I suppose. Run through and get me a drink of water son will you.

Uncle Archie could make people laugh at breakfast, even Danny's father — but still he had to go to work. He said, If you'd told me you were coming I could've made arrangements.

Ach, I was wanting to surprise yous all. Uncle Archie grinned: You'll be coming to the match afterwards though eh?

The father looked at him.

The boys're through at Airdrie the day.

Aw aye, aye. The father nodded, then he shrugged. If you'd told me earlier Archie — by the time I'm finished work and that...

Uncle Archie was smiling: Come on, long time since we went to a match the gether. And you're rare and handy for a train here as well.

Aye I know that but eh; the father hesitated. He glanced at the other faces round the table. He said, Naw Archie. I'll have to be going to my work and that, the gaffer asked me in specially. And I dont like knocking him back, you know how it is.

Ach, come on —

Honest, and by the time I finish it'll be too late. Take the boys but. Danny — Danny'll go anywhere for a game.

Uncle Archie nodded for a moment. How about it lads?

Not me, replied Danny's brother. I've got to go up the town.

Well then...Uncle Archie paused and smiled: Me and you Danny boy, eh!

Aye Uncle Archie. Smashing.

Here! — I thought you played the game yourself on Saturdays?

No, the father said, I mean aye — but it's just the mornings he plays, eh Danny?

Aye. Aw that'll be great Uncle Archie. I've never been to Broomfield.

It's no a bad wee park.

Danny noticed his mother was looking across the table at his father while she rose to tidy away the breakfast stuff. He got up and went to collect his football gear from the room. The father also got up, he pulled on his working coat and picked his parcel of sandwiches from the top of the sideboard. When the mother returned from the kitchen he kissed her on the cheek and said he would be home about half past two, and added: See you when you get back Archie. Hope the game goes the right way.

No fear of that! We'll probably take five off them. Uncle Archie grinned, You'll be kicking yourself for no coming — best team we've had in years.

Ach well, Danny'll tell me all about it. Okay then...he turned to leave. Cheerio everybody.

The outside door closed. Uncle Archie remained by himself at the table. After a moment the mother brought him an ashtray and lifted the saucer he had been using in its stead. He said, Sorry Betty.

You're smoking too heavy.

I know. I'm trying to...He stopped; Danny had come in carrying a tin of black polish and a brush, his football-boots beneath his arm. As he laid the things in front of the fireplace he asked: You seen my jersey mum?

It's where it should be.

The bottom drawer?

She looked at him. He had sat down on the carpet and was

taking the lid off the tin of black polish. She waited until he placed an old newspaper under the things, before leaving the room.

Hey Danny, called the Uncle. You needing any supporters this morning?

Supporters?

Aye, I'm a hell of a good shouter you know. Eh, wanting me along?

Well...

What's up? Uncle Archie grinned.

Glancing up from the book he was reading Danny's brother snorted; He doesnt play any good when people's watching.

Rubbish, cried Danny, it's not that at all. It's just that — the car Uncle Archie, see we go in the teacher's car and there's hardly any space.

With eleven players and the driver! Uncle Archie laughed: I'm no surprised.

But I'll be back in plenty of time for the match, he said as he began brushing the first boot.

Aye well you better because I'll be off my mark at half twelve pronto. Mind now.

Aye.

It's yes, said the mother while coming into the room, she was carrying two cups of fresh tea for herself and Uncle Archie.

Danny was a bit embarrassed, walking with his uncle along the road, and over the big hill leading out from the housing scheme, down towards the railway station in Old Drumchapel. But he met nobody. And there was nothing wrong with the scarf his uncle was wearing, it just looked strange at first, the blue and white, really different from the Rangers' blue. But supporters of a team were entitled to wear its colours. It was better once the train had stopped at Queen Street Station. Danny was surprised to see so many of them all getting on, and hearing their accents. In Airdrie Uncle Archie became surrounded by a big group of them, all laughing and joking. They were passing round a bottle and opening cans of beer.

Hey Danny boy come here a minute! Uncle Archie reached out to grip him by the shoulder, taking him into the middle of the group. See this yin, he was saying: he'll be playing for Rangers in next to no time...The men stared down at him. Aye, went on his uncle, scored two for the school this morning. Man of the Match.

That a fact son? called a man.

Danny reddened.

You're joking! cried Uncle Archie. Bloody ref chalked another three off him for offside! Eh Danny?

Danny was trying to free himself from the grip, to get out of the group.

Another man was chuckling: Ah well son you can forget all about the Rangers this afternoon.

Aye you'll be seeing a *team* the day, grunted an old man who was wearing a bunnet with blue and white checks.

Being in Broomfield Park reminded him of the few occasions he had been inside Hampden watching the Scottish Schoolboys. Hollow kind of air. People standing miles away could be heard talking to each other, the same with the actual players, you could hear them grunting and calling out names. There was a wee box of a Stand that looked like it was balancing on stilts.

The halftime score was one goal apiece. Uncle Archie brought him a bovril and a hot pie soaked in the watery brown sauce. A rare game son eh? he said.

Aye, and the best view I've ever had too.

Eat your pie.

The match had ended in a two all draw. As they left the terracing he tagged along behind the group Uncle Archie was walking in. He hung about gazing into shop windows when the game was being discussed, not too far from the station. His uncle was very much involved in the chat and after a time he came to where Danny stood. Listen, he said, pointing across and along the road. See that cafe son? Eh, that cafe down there? Here, half a quid for you — away and buy yourself a drink of ginger and a bar of chocolate or something.

Danny nodded.

And I'll come and get you in a minute.

He took the money.

I'm just nipping in for a pint with the lads...

Have I to spend it all?

The lot. Uncle Archie grinned.

I'll get chips then, said Danny, but I'll go straight into the cafe and get a cup of tea after, okay?

Fair enough Danny boy fair enough. And I'll come and get you in fifteen minutes pronto. Mind and wait till I come now.

Danny nodded.

He was sitting with an empty cup for ages and the waitress was looking at him. She hovered about at his table till finally she snatched the cup out of his hands. So far he had spent twenty five pence and he was spending no more. The remaining money was for school through the week. Out from the cafe he crossed the road, along to the pub. Whenever the door opened he peered inside. Soon he could spot his uncle, sitting at a long table, surrounded by a lot of men from the match. But it was impossible to catch his attention, and each time he tried to keep the door open a man seated just inside was kicking it shut.

He wandered along to the station, and back again, continuing on in the opposite direction; he was careful to look round every so often. Then in the doorway of the close next to the pub he lowered himself to sit on his heels. But when the next man was entering the pub Danny was onto his feet and in behind him, keeping to the rear of the man's flapping coat tails.

You ready yet Uncle Archie?

Christ Almighty look who's here.

The woman's closing the cafe.

Uncle Archie had turned to the man sitting beside him: It's the brother's boy.

Aw, the man nodded.

What's up son?

It's shut, the cafe.

Just a tick, replied Uncle Archie. He lifted the small tumbler

24

to his lips, indicated the pint glass of beer in front of him on the table. Soon as I finish that we'll be away son. Okay? I'll be out in a minute.

The foot had stretched out and booted the door shut behind him. He lowered himself onto his heels again. He was gazing at an empty cigarette packet, it was being turned in abrupt movements by the draught coming in the close. He wished he could get a pair of wide trousers. The mother and father were against them. He was lucky to get wearing long trousers at all. The father was having to wear short trousers and he was in his last year at school, just about ready to start serving his time at the trade. Boys nowadays were going to regret it for the rest of their days because they were being forced into long trousers before they needed to. Wide trousers. He wasnt bothered if he couldnt get the ones with the pockets down the sideseams, the ordinary ones would do.

The door of the pub swung open as a man came out and passed by the close. Danny was at the door. A hot draught of blue air and the smells of the drink, the whirr of the voices, reds and whites and blues and whites all laughing and swearing and chapping at dominoes.

He walked to the chip shop.

Ten number tens and a book of matches Mrs, for my da.

The woman gave him the cigarettes. When she gave his change he counted it slowly, he said: Much are your chips?

Same as last time.

Will you give us a milky-way, he asked.

He ate half of the chocolate and covered the rest with the wrapping, stuck it into his pocket. He smoked a cigarette; he got to his feet when he had tossed it away down the close.

Edging the door ajar he could see Uncle Archie still at the table. The beer was the same size as the last time. The small tumbler was going back to his lips. Danny sidled his way into the pub, but once inside he went quickly to the long table. He was holding the torn-in-half tickets for the return journey home, clenched in his right hand. He barged a way in between two men and put one of the tickets down on the table quite near to the beer glass.

I'm away now Uncle Archie.

What's up Danny boy?

Nothing. I'm just away home...He turned to go then said loudly: But I'll no tell my mother.

He pushed out through the men. He had to get out, Uncle Archie called after him but on he strode sidestepping his way beyond the crowded bar area.

Twenty minutes before the train would leave. In the waiting room he sat by the door and watched for any sign of his uncle. It was quite quiet in the station, considering there had been a game during the afternoon. He found an empty section in a compartment of the train, closed the door and all of the windows, and opened the cigarette packet. The automatic doors shut. He stared back the way until the train had entered a bend in the track then stretched out, reaching his feet over onto the seat opposite. He closed his eyes. But had to open them immediately. He sat up straight, he dropped the cigarette on the floor and then lifted it up and opened the window to throw it out; he shut the window and sat down, resting his head on the back of the seat, he gazed at the floor. The train crashed on beneath the first bridge.

F*lotsam and Salvage*

One day I was crabbing by myself on the beach. I had taken the kitchen bucket out with me and was absorbed in filling it with whatever creatures I could find at the tide's edge.

To turn over stones, to discover the secret life struggling there, was what I liked best to do. Shore crabs especially delighted me, their variety, the patterning on their backs, the range of colour from palest pink to plushy green, their size and ferocity. My pleasure in their appearance was adulterated by greedy acquisitiveness. I liked to have them, to keep them, to accumulate dozens of them in jars that grew green and smelly behind the garden gate.

But sometimes I would hold just one of them on the back of my hand and watch it, the extreme fragility of its hinged legs, the fringe of reddish hairs where they tapered, the frothy bubbling round the eye-stalks. The best were the tiniest, so small that like angels they could gather on a pinhead, so pale that it was hard to believe colour could ebb away to such paleness and still exist. Their shells were so thin that they had to be picked up with anguished delicacy in case a rough thumb-pad crushed through to whatever merest sliver of life quivered within.

Behind me a voice said, 'See's a shot o' yer pail.'

It was one of those upturning optimistic voices, not expecting anything but always hoping. They boy had come up quietly on plimsolled feet and was standing watching me.

'Whit're ye gettin' them fer? They're nae use.'

I had never thought of my crabs having a use. It was a new idea and I looked at him blankly.

'See's yer pail?' he said again, holding out a hand for it this time.

'What for?' I said.

'There's a wifie gies ye a thruppenny bit fer a bucket o' whelks,' he said.

He seemed to think that was answer enough. No one who had a bucket would spend their time gathering useless crabs when you could fill it with whelks and get threepence for it. I saw the logic of that, which didn't mean that I was going to hand over my bucket to him.

'Can anybody do it?'

Perhaps you had to be one of the fisher families before you could earn your threepenny.

'Onybody wha hes a pail,' he answered flatly.

'There's another pail in our shed.'

'Well, awa' and get it, then. Leave that ane here.'

I did what he told me, partly because I was quite unaccustomed to being told what to do by ragged boys. I knew it was only my bucket he wanted, but he seemed to be offering a kind of friendship in exchange for it.

Our house had a garden, and a back gate. I was very careful not to be seen taking the coal-bucket out of the shed in case somebody called me in, to dust chairs or go shopping. It was a heavy galvanized bucket with a clanking handle.

When I got back to the beach the boy had already tipped out all my crabs and begun to gather whelks. He was quite efficient, but I realized fairly quickly that it was a game I could win. I had spent so many hours alone on this beach that I knew all its contours, all its surface topography. There were certain pools where you could count on finding whelks in large colonies, others where whelks never showed. The beach had its own social communities, as rigidly structured as the town's. Where the mussels flourished, the whelks did not, because the thickly clustering mussel-nurseries left no toehold for wandering whelks. Some pools were so constructed that the tide scoured them too thoroughly for any creature but the limpet to be left holding on. Some others were not often enough reached by the tide and grew rusty and too stagnant for even the scavenging whelks to thrive. In some, jelly-bright sea-anemones menaced any life that passed their tentacles. But there was one pocket in the farther-out rocks which I knew to be rich in whelks. It was a deep round hole ground out by eddying tides under an overhang of rock that was uncovered for only a short time each day. Weed grew thickly there, and whelks clung to every frond and bladder of wrack, swarming over the barnacled rock, falling in piles to the bottom of the hole only to begin the climb again to the air.

The boy was wandering in an unplanned way, picking up a few shells here, a few there. He had covered already quite a large area of beach, but I could see that he had by no means picked it clean. His bucket was about a quarter filled but it would take him, I knew, a lot longer to fill it to the top.

I left him and ran out to the farthest reach of the tide, where the thick stems of sea-wrack were anchored to the rocks, jumping lightly among the yellow slithering blades of weed until I reached the deep hole. Then I pushed up my jersey sleeves, lay on my front and dipped my arm down into the water. I brought up whelks in handfuls, working as if I were driven, as indeed longing to earn that threepenny bit did drive me. When I felt bare rock at the bottom of the pool, the shifting mass of shells all transferred to my bucket, I sat up and began to pick more of them off the rock. I lifted the heavy strands of weed to uncover more and more whelks hiding underneath. In perhaps quarter of an hour my bucket was brimming over. I got to my feet carefully and began the walk back. The bucket was very heavy now and it pulled uncomfortably at my shoulder muscles.

The boy came to meet me. His own bucket was still light, only half-filled, but he showed no hint of surprise at my success.

'Ye've got ower mony' was all he said.

He took my bucket from me and tipped the top layers of whelks into his own so that both were respectably full but neither of them spilling over.

'She'll no gie ye mair than a thruppenny,' he said, making it clear that he wasn't stealing my labour but, on the contrary, sharing his expertise with me.

'C'mon,' he said, and I followed him, the bucket easier now, over the beach and up the stone steps to the esplanade.

The old woman's shed was in a side-street leading up from the beach. There was a row of fishermen's cottages, built so low that I could have touched the stone-slated roofs without stretching. Halfway along the row was a close entrance, dark, narrow and smelly. The boy led the way up the close and into a backyard lined all round with wooden sheds. The walls of the sheds were shiny black with layers upon ancient layers of tarring. One of the shed doors was open, and an old woman sat in the doorway behind a battered kitchen-table. On the table were piles of pink paper and a tobacco-tin full of pins. She was putting the pins carefully, in neat rows, all the heads pointing the same way, into cut-up pieces of pink paper. She wore a huge black skirt, so long and full and thick that she seemed to be supported on it, and over it a flowered apron, equally voluminous. There was a crocheted shawl wrapped and pinned round her shoulders and a man's flat cap on her head. I would never have dared to speak to her on my own, but the boy seemed to have her measure.

'Well, Jeemie, is this you efter sweetie money?' she asked, ignoring me.

Jeemie put his bucket on the table in front of her and, seeing that I didn't move, took mine from my hand and lifted it up beside his own. She looked at them without interest.

'They'll be a' steens,' she said.

'They're nae. There's nivir a steen. They're a' buckies.' The boy was bitter.

'Well, well,' she said. 'We'll need tae see.'

She got up with surprising ease, lifted each bucket and tipped the whelks into a bin behind her, watching carefully as they fell. 'Right enough,' she said, then 'They're buckies right enough. Yer wee pal mun be a better worker nor you.'

She scrabbled behind her apron, into the folds of her skirt, bending sideways to reach down into some inner pocket, and came up with two silver threepenny bits laid flat on her hard black palm.

The boy took them both from her, looked at them as if there might be some difference between them, and then passed one over to me. I was nine years old and I had never earned money before. It made me bold.

'Can we get you some more, please?' I asked.

'Oh ay,' she said, almost smiling. 'If ye're nae feart o' work, ye can come back, lassie, fetch old Bess mair whelks. I canna bend fer them masel'.'

'What does she do with them?' I asked Jeemie when we got back on the road with our empty buckets.

'Whit wid she dae wi' them? She sells them.'

This time I led the way back to the beach and out to the tide's edge. Jeemie followed me out to the rocky part, admitting, without saying so, that I was good at finding the best places even if I knew nothing about commerce. We didn't talk. There seemed no need. Our purpose was quite clear and required no discussion. It was simply understood between us that we were to become the champion whelk-gatherers, unsurpassed for speed and quality, on our way to making a fortune in threepenny bits. Heads down, hands flying, we flung the whorled, horny-doored whelks into our buckets pitilessly. I, who had spent long minutes just watching a whelk travel across a rock, its flanged foot smoothing silvered glue behind it, could now only grab and gather, grab and gather. In no time we had two more buckets filled and were on our way back to Bess's shed.

Her door was shut, the table pushed outside, its papers of pins cleared away.

'She's awa' fer her denner, the auld bitch,' said Jeemie, putting down his bucket.

I thought that was the end of it, but the boy went across the yard to a kitchen door and knocked on it. There was a long wait. He knocked again. No answer. He opened the door and shouted in: 'Hey! We've brocht yer whelks.'

Grumbling, she came to the door and looked at us.

'Come back when I've had ma denner. Ye can leave yer pails by the shed door.'

Jeemie stood his ground.

'We need oor pails back.'

She disappeared into the kitchen for a moment and came out again with a huge iron key. It was a key fit for a dungeon, for a medieval prison, for a baron's keep. She opened the tarred shed-door and stood angrily waiting for us to carry the buckets to her. She snatched them from us, tipped the whelks into her bin without a word and handed back the emptied buckets. Then she locked the shed again and stood waiting for us to go away.

Jeemie held out his hand, palm upwards, matching her silence with his.

'Oh!' said she. 'Ye want paid dae ye!'

Her hand went down, down into her skirt pocket, fumbled there a minute and came up with a sixpence. She made to put it in Jeemie's palm, then snatched it back.

'Na, na. We'll gie it tae the lassie. It was her did the work, I bet.' And she offered the sixpence to me. 'And dinna come back. I'm nae needin' ony mair whelks the day.'

We went out of the yard and stopped.

'You give me your threepenny and I'll give you the sixpence,' I said.

'Na,' he said. '*You* keep the saxpence and gie us the thruppenny.'

'Why?' I thought there must be some catch to it.

'Because if ma mither sees saxpence she'll tak it aff o' us. If I hev twa thruppennies I can gie her ane and keep ane.'

It was another world. I didn't argue. I tied the sixpence in a knot in the corner of my hankie, put the hankie in my elasticated knicker-leg, and gave him the threepenny bit. We wandered, slowly now, back to the beach. It seemed to be taken for granted that we should spend the rest of the morning together even if there was now no fortune to be made.

The tide had turned. The best whelk-pools were already covered. Jeemie put down his bucket and began to skiff flat stones across the surface of the water — three, four, five, six times he could make his stone jump before it sank. There was a lot of driftwood draggling on the tide. I wandered along the edge, aiming stones at floating sticks, until I saw something bigger drifting in. It looked like a small wooden barrel. I shouted for Jeemie and together we began to bring it ashore, throwing stones as big as we could lift to fall just beyond the barrel so that the splashes drove it closer. It hung low in the water, as if there were something heavy inside. 'Brandy for the parson, baccy for the clerk'? The poem floated into my mind, but I kept quiet. Jeemie wasn't the sort of boy for poetry.

When it was very close we splashed in and brought it ashore. It was a small, tightly sealed keg, hooped with iron and very heavy. Between us we rolled it to the top of the beach and flopped there a moment, puffing.

But there were more. All along the tide's edge now we could see that each successive wave was bringing in pieces of broken boxes and barrels.

'There was a ship went doon last night,' said Jeemie sudenly. 'Oot at the bar. A Jerry plane got her.

We ran back to the edge of the water and began labouring to bring in the wreckage. There was one more whole barrel, a few empty wooden crates and lots of sections of broken barrel, some with the contents still adhering.

It was butter. Bright-yellow, glossy New Zealand butter, the cold North Sea water keeping it hard and fresh. Butter in such quantities as neither Jeemie nor I had seen. Butter was kept under the counter at the local grocer's, doled out in tiny pats in exchange for ration coupons. My mother used to mix it with margarine to make it go farther. She would probably go mad at the sight of all that butter wasting on the waves. The tide was beginning to deposit it on the shore and the stones were becoming slithery with it.

We realised that we must get as much of it home as possible before everyone else knew about it. We rescued the second whole barrel and rolled it up beside the first. I took off my old school coat and draped it over them. Jeemie had no coat. Then we took the buckets down to the line of driftwood and began to fill them with the best chunks of butter, breaking away the barrel spars to find the cleanest parts. It took only a few minutes to fill a bucket each.

'You can take that bucket home,' I said to Jeemie. 'But you must bring it back when your mother's emptied it.'

'Na, I'll jist tak a bitty,' he said. 'They'll niver believe I havena pinched it.'

I looked at him. I could see the difficulty. He was exactly the sort of boy people did accuse of stealing things. Not very clean, a bit shifty, too quick on his feet, as if he'd been dodging a belt on the ear all his life.

'I'll help you carry it up to your place,' he said.

My house was on the esplanade, right beside the beach. We had only to cross the road to be facing our gate. Jeemie and I walked together until we reached it and then I dumped my bucket and rushed in, screaming the news about the butter as I ran.

My mother came out of the kitchen, telling me to be quiet, but curious about the cause of my hubbub. When she saw the two buckets of butter standing on her doorstep, when I told her that there were barrels full of it still on the beach, when she had tested it with her finger, smelled it and tasted it, she almost screamed herself. I could see the scream bubbling up in her throat. She gagged on it, choked, couldn't find words. Then she looked up and down the street anxiously, lifted up the buckets, brought them inside and shut the door.

'Did anyone see you?' she asked. 'We'll have to get it put away quickly. Here, help me.'

She tipped the contents of the buckets into a basin in the sink, filled it with cold water and washed the chunks of butter clear of the grit and weed that clung to it. Then she went to the drawer where she hoarded string and paper and brought out squares of carefully folded, used greaseproof paper. She cut the butter into convenient pieces and showed me how to wrap each piece neatly in paper. I didn't understand the need for hurry but I enjoyed the feeling of richness, of having such a quantity of something. The contrast between this abundance and the usual ration was very exciting. I could feel that my mother was excited, too, although she hardly said a word, just worked and worked to get the butter wrapped up and put away. She stowed it all to the very back of the larder, washed out the buckets and put them away.

'Now,' she said. 'Did you say there was a whole barrel still on the beach?'

'Two,' I said. 'We hid them up at the top, near the steps.'

Then suddenly I remembered Jeemie and ran to the door. Of course he had gone. I felt strange. I had left him on the doorstep without a word, forgotten all about him while I helped my mother. I knew that it was hopeless, that I could not have brought him in, cleaned him up, put him away like the butter. But it was a sad sort of feeling.

My mother had brought the old pram out of the shed and was hurrying with it across the road to the beach. I hurried after her, to show her where Jeemie and I had hidden the kegs.

Already there were people on the beach, gathering up pieces of butter as they washed in with the tide. I had never seen my mother being furtive before. But now, after the long years of rationing and mean eating, the contriving and eking out, to be faced with butter in such quantities had tipped her over some edge of conventional behaviour. She wanted that butter more than she wanted respectability.

All the other people were at the foot of the beach, at the edge of the water, catching the pieces of floating cargo as it came ashore. It was not too difficult for us to haul out our cache from under the steps and heave it into the pram without being seen.

'Now fetch some wood.'

My mother began to put pieces of driftwood on top of the pram, hiding the two kegs of butter completely. Then, quite boldly, she pushed the load back to the garden gate and slid the whole thing into the coalshed. A fine load of firewood, it looked like.

People were beginning to scurry off the beach back to their homes, some carrying tin basins full of butter, some holding apronsful in front of them. Those who came last could find only scrapings on the rocks, or lumps of butter that had gathered too much detritus as it rolled in with the tide and was now full of splintered barrel-wood and grit. But every scrap was gathered in.

Later in the day, when the Customs men came down to our part of town, investigating a rumour that a cargo of butter had been washed ashore from the wrecked ship, there was nothing to see. All the butter had been spirited away, all the barrel staves gathered for firewood, and no one knew anything.

That night, when I was undressing for bed, I found the knotted handkerchief in my knicker leg, the sixpence still safely inside. I put it in a safe place, to spend on some much duller day.

NEIL GUNN

M*orning Tide*

The boy's eyes opened in wonder at the quantity of sea-tangle, at the breadth of the swath which curved with the curving beach on either hand. The tide was at low ebb and the sea quiet except for a restless seeking among the dark boulders. But though it was the sea after a storm it was still sullen and inclined to smooth and lick itself, like a black dog bent over its paws; as many black dogs as there were boulders; black sea-animals, their heads bent and hidden, licking their paws in the dying evening light down by the secret water's edge. When he stepped on the ware, it slithered under him like a living hide. He was fascinated by the brown tangled bed, the eel-like forms, the gauzy webs. There had been no sun to congeal what was still glistening and fresh.

A faint excitement touched his breast; his lips parted; his eyes shot hither and thither. He began rooting at the bed with his boots, stooping every now and again to examine the head of a tangle. At length he found one with a small delicate limpet stuck in the cup of its head; a young one because its round stem was slim and not two feet in length. As he snicked it free its leathery tail-frond flicked sea-mist to his face. His teeth began to water. He cut the brown stem two inches from the shell with his pocket knife, which had one strong sharp blade. As he pared off the claw-roots that curled round the hollow where the shell was, he was very careful not to remove the shell. The shell was the jewel in its head, importing tender-ness and sweetness. It was also something to 'show off'. He put the tangle head in his pocket and lifted the folded sack that had slipped from under his arm. Then he went on rooting amongst the bed until he had found two more tangles with shells. But neither was so delicate, so thin-skinned, as the first. The first was a beauty! When he had dressed them and stuffed them in his pocket, he gave an involuntary shiver; his teeth clicked and he brought the back of his hand to his wet nose. His nose was colder than his hand. His body twisted and wriggled inside his clothes searching for warmth. But his round dark eyes were on the boulders down by the hidden sea-edge. It was time he got his 'baiting'.

With a final look at the shredded end, he stowed the tangle in his pocket. If he didn't hurry it would soon be dark. And his father would be waiting for the bait. And, anyway, he would pick up another young tangle with a shell on his way back. He would keep it whole to show off, because the other boys were playing football. He himself liked playing football better than anything. First of all he had said to his mother that he wouldn't go to the ebb. The mood had come on him to be dour and stubborn. It was a shame that he should have to go to the ebb, with the other boys playing football. The injustice of it had hit him strongly. His father had said nothing. But if he didn't go then his father would have had to. And by two or three in the morning his father would have to be aboard. It must be terribly cold in the dark out on that sea; out on that sea in the small hours of a sleety morning, or on a morning of hard frost, with a grey haar coming off the water like ghosts' breath.

The loneliness of the bouldered beach suddenly caught him in an odd way. A small shiver went over his back. The dark undulating water rose from him to a horizon so far away that it was vague and lost.

A short distance away, right on the sea's edge, he saw one of the boulders move. His heart came into his throat. Yet half his mind knew that it could only be some other lonely human in the ebb. And presently he saw the back bob up for a moment again.

Yes, it was a man. Seeking among the boulders there like some queer animal! He looked about him carefully. There was no one else. There were just the two of them in the ebb. Here they were on this dark beach, with nobody else. A strange air of remoteness touched him. It was as though they shared this gloomy shore, beyond the world's rim, between them. There was a secret importance in it.

As he came erect, the man who was also gathering mussels in the ebb straightened himself and put his hand in the small of his back. His back curved inward. So it was Sandy Sutherland. Swinging his bag onto a boulder, the boy got under his burden and started off towards Sandy, but yet slanting away from him a little to make his approach not too deliberate.

'Is it yourself, Hugh, that's in it?' The note of surprise in Sandy's voice was pleasant and warm. His face was whiskered, his ways quiet, his eyes dark-shining and kind.

'Yes,' said Hugh, pausing, and tingling a little with shy pleasure.

'Getting your father's baiting, I see.'

'Yes.'

'Aren't you cold?'

'Oh no.'

'Well, you're plucky,' said Sandy. 'And it's your father that's lucky. Here I have to be down myself.' He said this with no grudge in his voice. It was half a joke.

'Well, I'll be coming with you, if you wait till I gather one or two more. What sort of baiting have you yourself?'

'Oh, I think I have plenty.' Hugh dropped his bag on a boulder and caught Sandy's eyes measuring the quantity in a secret glance.

'Perhaps you have, then.'

'Oh, I think so,' murmured Hugh.

But there was a faint reserve in Sandy's voice, as though he didn't like to suggest that the boy might not have gathered a full baiting.

It was a delicate point. Hugh saw that if Sandy could find a hidden or off-hand way of adding a few, he would do so.

'Perhaps you have, then,' Sandy repeated. 'Though you could gather one or two more for luck if you like while I'm finishing off. It would save you to-morrow night maybe.'

'Father isn't wanting a full baiting,' Hugh explained. 'There's not much short of half one at home.'

'Oh, that makes a difference! To gather more would be waste. Indeed, I'm thinking you have a full baiting there as it is, if not more.' The voice was amused rather than relieved. But Hugh understood perfectly.

'I think there's plenty,' he considered, as if now a trifle doubtful.

'Plenty? I should say so! Do you think I have as many as you myself?' He measured both bags with an eye. 'I should say nearly.'

'You have more, I think,' judged Hugh, politely.

'Well, if you say so. But remember it's you I'll blame if I'm short! It's a cold place this, anyway.' He lit his pipe, sending great clouds of blue smoke about his head. 'We'll chance it. So let us be going.' He shouldered his bag cheerfully and they both set off.

As the man and the boy went on they spoke together companionably. The creek or harbour was on their right hand. Between them and it were three boats that had been hauled up and lashed with ropes to driven piles. No highest storm could touch them there. When a boat is hauled to its last mooring it gives up its spirit, and its head-rope becomes a sagging halter.

On their left hand, in the middle of the brae-face, was a line of thatched cottages. Yellow lights gleamed in the little square windows. There was still sufficient daylight left for the white-washed walls to gleam also, but with a whiteness that came against the eyes in a ghostly way.

'Is the bag heavy on you?' asked the man.

'Oh no,' said the boy.

'You'll be going to school every day?'

'Yes.'

'How old are you now?'

'I'll be thirteen next birthday.'

Sandy almost stopped.

'Go' bless me, you're a big boy for your age!

Hugh smiled shyly. He was tall for his age but thin, and could if necessary twist his legs like a tailor. His elbows and his knees did not stick out, and he could run on his toes. Only sometimes, and privately, did he feel himself too light, did he know himself less weighty than one or two boys of his own age, a something light and airy that could be bruised with a stocky fist and knocked off its balance. Sometimes he regretted this very much, and longed for a fist that would sheer through hands and arms as through the leaves and twigs of a tree. But only for moments, his body was so lithe.

But now as he walked along the road with Sandy he might

well enough be a man going to sea. Some of the importance of this manliness set him apart from boyish things for the moment. Their voices were reasonable and unhurried. The night set them apart. The bag on his back, the bag on Sandy's back, their hands gripping over their breasts, walking on, talking to each other with a quiet confidence, their heads stooping forward, the two of them.

'Well, I'll leave you here,' and Sandy drew up.

Hugh felt a little shy.

'How's your mother — how are they all at home?'

'Very well, thank you.' Hugh wanted to ask how Mrs Sutherland was, but felt the question too important for one of his age.

'Good-night, then, Hugh. Tell your father I'll be looking for him in the morning.'

'I will,' Hugh answered readily. 'Good-night.'

'If I come down past your way, I'll throw a stone at his window, tell him,' called Sandy out of the gloom.

'All right!' called Hugh, laughter in his tone. As he went on alone he felt confident and happy. For a few steps, he ran, but the bag bumped and slewed round, nearly choking him. He would drop it presently when no one would see that he had to take a rest.

With loud chatter, some boys loomed round a bend. The Seabrae boys going home from football. Hugh walked calmly, like a man.

'Ha! ha! here's Hugh!' cried Rid Jock, son of Magnus.

The other boys also cried the same thing.

Hugh's manner was quiet. 'Coming home from the football?' he asked.

'Why didn't you come? You said you'd come in the morning.'

'I said I might,' corrected Hugh mildly.

'You didn't. You said you would,' corrected Jock positively.

'Yes, I heard you,' piped little Dannie. And the others piped the same thing. 'I heard you say it! I heard you say it!'

'You didn't,' said Hugh.

'Am I a liar?' asked Jock.

'You are, if you say that,' said Hugh.

'Oh I am, am I?' asked Jock. 'Ho! ho! I'm a liar, am I?'

'You are,' repeated Hugh, who had not yet raised his voice.

'Say that again,' threatened Jock loudly.

Hugh, ignoring him, took a step on. Immediately his bag was tugged from behind. He swung round. Dannie scampered off, and from a safe distance shouted, 'Does your mother know you're out?'

The other boys laughed loudly as they backed a step or two.

'His mother wouldn't let him to the football — in case he would fall!' All the boys laughed and jeered, repeating the fatal insult in sing-song. Hugh made no move. This inaction at once frightened and goaded them.

It was nothing that this very morning at school he had been on easy terms with them all. A roving homeward band had happened on a solitary doing something different from themselves, and their minds immediately became excited and cruel. Moreover, this solitary did not belong to Seabrae. He was a bee from another hive, a gull from another rock.

'Mamma's pet!' shouted a voice, and encouraged by the sally, Jock suddenly hit Hugh's unprotected face.

In the same instant the bag dropped and Hugh sprang at him. Self-consciousness got blotted out in fury. He saw nothing but the blur of Jock's face and lashed at it. Nothing would stop him so long as he could see and stand. His fury was destructive and primal. Blows rained on his own face. They merely made his fury red. Something had to give way. It was Jock, because Jock, though a good fighter, could not be possessed by a demon.

The little band with their leader broke and fled, shouting insults. Hugh leapt after Jock and tripped him as he ran, so that Jock fell headlong on his face. Hugh sprang on his back. Jock covered up his face and squawked his fear and rage. Hugh, pinning down the neck with his left hand, walloped the head with his right, shouting, 'Get up! get up!'

Some of the other boys came back.

'Get off him!' they yelled threateningly. He ignored them. One boy tried to kick him. He leapt from Jock's back. A stone hit him on the breast. All the boys, including Jock, retreated. Stones fell around him where he stood stock still on the middle of the road, quivering.

'Never mind; you blooded his nose!' said a consoling voice

in the distance to Jock, whose face was damaged more by the gravel of the road than by Hugh's fists.

Automatically Hugh's fingers came up to his nose and touched there the slime of blood. He peered at the reddened tips. He now felt the blood running down his upper lip. It was also trying to slide down the back of his throat. The taste of it was salty and sticky in his mouth. Drops fell on the back of his hand. They had been falling upon the breast of his jersey. He pressed his knuckles against his nose, and his throat swallowed a lot of the blood before he could stop it. He held his head back, and then the blood went down his throat freely. You always have to hold your head back when your nose is bleeding. It's the only way to stop it. But you should always at the same time hold something cold to the back of your neck.

He stumbled over the edge of the road holding his head back, over the grassy bank and down to the edge of the river or burn. He was not thinking of anything but the blood. All his flesh was trembling. He was not really frightened of the blood. Here he was with his nose bleeding after a fight, and he was calmly stretching backwards against the bank and pressing a cold stone to the nape of his neck.

He lay quite still, his eyes to the pale stars. There were not many; one here and there. The sky was very very far away. It was quiet, too. It was high and dark, and the stars were lonely in it. He had to swallow every now and then. The blood was stopping. It was strange and lonely on your back, looking at the stars, down by the burn. He was doing all this by himself. He had fought Jock — the whole of them — and sent them flying. His teeth closed and his face went cold. His face felt like a cold wedge and his eyes glittered relentlessly.

One could weep through sheer rage. But not even that would he do again. No matter what happened to him in the world, never would he cry more. He had conquered — and sent them flying. Everyone knew that the fight with Jock had to take place, and now it was all over and he was lying here. His nose had been blooded, but — that often happened at a fight. He had seen it happen once with grown men. But when a grown man had his nose blooded, he did not stop fighting for that. Only boys stopped fighting when they saw blood.

JANICE GALLOWAY

Fearless

There would be days when you didn't see him and then days
when you did. He just appeared suddenly, shouting threats
up the main street then went away again. You didn't question
it. Nobody said anything to Fearless. You just averted your
eyes when he was there and laughed about him when he
wasn't. Behind his back. It was what you did.

Fearless was a very wee man in a greasy gaberdine coat meant
for a much bigger specimen altogether. Greygreen sleeves
dripped over permanent fists so just a row of yellow knuckles,
like stained teeth, showed below the cuffs. One of these fisted
hands carried a black, waxed canvas bag with an inept burst
up one seam. He had a gammy leg as well, so every second
step, the bag clinked, a noise like a rusty tap, regular as a
heartbeat.
  He wore a deceptively cheery bunnet like Paw Broon's over
an escape of raw, red neck that hinted a crewcut underneath,
but that would've meant he went to the barber's on a regular
basis, keeping his hair so short, and sat in like everybody else
waiting his turn, so it was hard to credit and since you never
saw him without the bunnet you never knew for sure. And he
had these terrible specs. Thick as the bottoms of milk bottles,
one lens patched with elastoplast. Sometimes his eyes looked
crossed through these terrible specs but it was hard to be sure
because you didn't get to look long enough to see. Fearless
wouldn't let you.

There was a general assumption he was a tramp. A lot of
people called him a tramp because he always wore the same
clothes and he was filthy but he wasn't a tramp. He had his
own house down the shore front scheme; big black finger-
stains round the keyhole and the curtains always shut. You

could see him sometimes, scrabbling at the door to get in, looking suspiciously over his shoulder while he was forcing the key to fit. There were usually dirty plates on the doorstep too. The old woman next door cooked his meals and laid them on the step because he wouldn't answer the door. He sometimes took them and he sometimes didn't. Depended on his mood. Either way, there were usually dirty plates. The council cut his grass, he had daffodils for christsake — he wasn't a tramp. He was the kind that got tramps a bad name: dirty, foulmouthed, violent and drunk. He was an alkie all right, but not a tramp: the two don't necessarily follow.

The thing about Fearless was that he lived in a state of permanent anger. And the thing he was angriest about was being looked at. Sometimes he called it MAKING A FOOL OF and nobody was allowed to get away with it. It was rule and he had to spend a lot of time making sure everybody knew it. He would storm up and down the main street, threatening, checking every face just in case they didn't know, then if he thought he'd caught you looking he would stop, stiffen and shout WHO ARE YOU TRYING TO MAKE A FOOL OF and attack. Sometimes he just attacked: depended on his mood. Your part was to work out what sort of mood it was and try to adjust to it, make the allowance. It was what you were supposed to do. Most folk obliged, too — went out of their way to avoid his maybe-squinty eyes or pointedly NOT LOOK when they heard the clink and drag, clink and drag, like Marley's ghost, coming up the street. Then the air would fall ominously silent while he stopped, checking out a suspicious back, re-inforcing his law. On a bad day, he would just attack anyway to be on the safe side. Just in case. You couldn't afford to get too secure. There was even a story about a mongrel stray he'd wound into a half-nelson because it didn't drop its gaze quick enough, but that was probably just a story. Funnier than the catalogue of petty scraps, blows that sometimes connected and sometimes didn't that made up the truth. It might have been true right enough but that wasn't the point. The point was you were supposed to laugh. You were meant to think he was funny. Fearless: the very name raised smiles and humorous expectations.

Women shouted their weans in at night with HERE'S FEAR-
LESS COMING, or squashed tantrums with the warning
YOU'LL END UP LIKE FEARLESS. Weans made caricatures
with hunchback shoulders, cross-eyes and a limp. Like
Richard the Third. A bogeyman. And men? I have to be careful
here. I belonged to the world of women and children on two
counts, so I never had access to their private thoughts voiced
in private places: the bookie's, the barber's, the pub. Maybe
they said things in there I can have no conception of. Some
may have thought he was a poor old soul who had gone to the
bad after his wife left him. Romantics. I suppose there were
some who could afford to be. Or maybe they saw him as an
embarrassment, a misfit, a joke. I don't know. What I do know
is that I never saw any of them shut him up when the anger
started or try and calm it down. I remember what women did:
leaving food on the doorstep and bottles for him to get money
on; I remember women shaking their heads as he went past
and keeping their eyes and their children low. But I don't
remember any men doing anything much at all. He didn't
seem to touch their lives in the same way. They let him get on
with what he did as his business. There was a kind of respect
for what he was, almost as if he had a right to hurl his fists,
spit, eff and blind — christ, some people seemed to admire
this drunken wee tragedy as a local hero. They called him *a
character. Fearless is a character right enough* they would say
and smile, a smile that accounted for boys being boys or something like that. Even polismen did it. And women who
wanted to be thought above the herd — one of the boys.

After all, you had to remember his wife had left him. It was
our fault really. So we had to put up with it the way we put up
with everything else that didn't make sense or wasn't fair; the
hard, volatile maleness of the whole West Coast Legend. You
felt it would have been shameful, disloyal even, to admit you
hated and feared it. So you kept quiet and turned your eyes
away.

It's hard to find the words for this even now. I certainly had
none then, when I was wee and Fearless was still alive and
rampaging. I would keek out at him from behind my mother's

coat, watching him limp and clink up the main street and not understand. He made me sick with fear and anger. I didn't understand why he was let fill the street with himself and his swearing. I didn't understand why people ignored him. Till one day the back he chose to stop and stare at was my mother's.

We were standing facing a shopwindow, her hand in mine, thick through two layers of winter gloves. The shopwindow was full of fireplaces. And Fearless was coming up the street. I could see him from the other end of the street, closer and closer, clinking the black bag and wheeling at irregular intervals seeing if he could catch somebody looking. The shouting was getting louder while we stood, looking in at these fireplaces. It's unlikely she was actually interested in fireplaces: she was just doing what she was supposed to do in the hope he'd leave us alone — and teaching me to do the same, I suppose. Fearless got closer. Then I saw his reflection in the glass: three days' growth, the bunnet, the taped-up specs. He had jerked round, right behind where we were standing and stopped. He looked at our backs for a long time, face contorted with indecision. What on earth did he think we were plotting, a woman and a wean in a pixie hat? What was it that threatened? But something did and he stared and stared, making us bide his time. I was hot and cold at once suddenly sick because I knew it was our turn, our day for Fearless. I closed my eyes. And it started. A lot of loud, jaggy words came out the black hole of his mouth. I didn't know the meanings but I felt their pressure. I knew they were bad. And I knew they were aimed at my mother. I turned slowly and looked: a reflex of outrage beyond my control. I was staring at him with my face blazing and I couldn't stop. Then I saw he was staring back with these pebble-glass eyes. The words had stopped. And I realised I was looking at Fearless.

There was a long second of panic, then something else did the thinking for me. All I saw was a flash of white sock with my foot attached, swinging out and battering into his shin. It must have hurt me more than it hurt him but I'm not all that clear on the details. The whole thing did not finish as heroically as I'd have liked. I remember Fearless limping away, clutching the ankle with his free hand and shouting about a liberty, and my mother shaking the living daylights out of me, a furious telling off, and a warning I'd be found dead strangled up a close one day and never to do anything like that again.

It was all a long time ago. My mother is dead, and so, surely, is Fearless. But I still hear something like him; the chink and drag from the closemouth in the dark, coming across open, derelict spaces at night, blustering at bus-stops where I have to wait alone. With every other woman, though we're still slow to admit it, I hear it, still trying to lay down the rules. It's more insistent now because we're less ready to comply, look away and know our place. And I still see men smiling and ignoring it because they don't give a damn. They don't need to. It's not their battle. But it was ours and it still is. I hear my mother too and the warning is never far away. But I never could take a telling.

The outrage is still strong, and I kick like a mule.

Boys among Rock Pools

Boys on knees,or prostrate, and scrambling
About rocks, by rock pools and inlets,
Noting with accurate eye the wash of water.
They hunt (O primitives!) for small fish,
Inches long only, and quicksilver,
But pink beneath the dorsal fin
Moving with superb locomotion.
Bodies bent, eyes all upon the prey —
Boys in shallow water with sun-warmed feet.

GEORGE MACKAY BROWN

Winter Bride

The three fishermen said to Jess of The Shore
'A wave took Jock
Between The Kist and The Sneuk.
We couldn't get him, however we placed the boat.
With all that drag and clutch and swell
He has maybe one in a hundred chances'
They left some mouthing cuithes in the door.
She had stood in this threshold, fire and innocence,
A winter bride.
Now she laid off her workaday shawl.
She put on the black.
(Girl and widow across a drowned wife
Laid wondering neck on neck.)
She took the soundless choir of fish
And a sharp knife
And went the hundred steps to the pool in the rock.
Give us this day our daily bread
She swilled and cut
And laid psalms and blessings on her dish.

In the bay the waves pursued their indifferent dances.

cuithes coal fish

49

PETER BUCHAN

Black Beasties that Bring Golden Reward

The lug worm is not a beautiful sight. It is a fat, black beastie which makes its home in the sandy bottoms of estuaries or sheltered beaches, where it betrays its presence by leaving little heaps of droppings on the sands around low-water mark.

If you dig a hole on the seaward side of these little heaps, you'll find the lug which, despite its appearance, makes a wonderful bait.

Now, to obtain a few lug for an evening's angling is not a major problem, but should you require several hundred to bait a fleet of small lines the problem becomes acute and you'll have to 'send awa'' for lug in bulk.

The best place to get lug is Ardersier, where extensive mudflats make a perfect lug nursery.

So you send a telegram to the mannie at Ardersier. Post-haste he will send you a biscuit tin full of lug which you must collect at Peterhead station where they will arrive on the last train at 8.50 p.m.

The cost will probably stagger you, for, including the telegram and the freight charges, your McVitie's tin of worms will cost no less than twenty-three shillings! How can you be expected to make a living with bait at such a price?

Well, that's your own problem, so you just have to carry the bait home where the lines are ready for baiting.

But I almost forgot that during the day you've been busy getting a bucket of limpets from the rocks.

The best tool for 'hacking' limpets is an old straight-backed table knife with the handle well bound in flannel to ensure a good grip. This tool you will call a 'sprod' and, like countless others before you were born, you'll make the fatal mistake

50

of holding your sprod like a dagger.

Sure, you'll dislodge the limpet at the first fell swipe but in so doing you'll skin your knuckles on the barnacles, which were invented for the express purpose of drawing blood.

You'll soon get the knack but it will take you a long time to fill a bucket of limpets, especially if several other folks are on the rocks after the same errand.

How to get the beasties out of their shells?

Plot them, you silly!

Pour boiling water over them and they'll fairly loup oot! But first you've to carry them home to join the lug.

Baiting the lines is a slow, painstaking job. You'll coil the line neatly into a wooden 'backet', laying the baited hooks in neat rows so that they will run clear during the shooting process. And between the rows you'll put strips of newspaper to keep the hooks apart.

It will take till nearly midnight for you and your mates to bait the six lines, comprising 1,200 hooks. The bait will be lug and limpet alternately, or 'time aboot', and when the job is finished you'll wash your stinking hands with Lysol before having a cup of tea.

Then it's off to sea to shoot the lines in the bitter cold of a winter morning. You'll be in a small boat whose only lights are paraffin lamps or the glare from a 'torch', which is like a kettle with a wide spout from which several strands of twisted wick protrude.

The fuel, of course, is paraffin. You'll have to be very careful because the boatie will do her best to pitch you overboard and there's not a great deal of room. You'll need to watch your fingers, too, because the lines are run out while the boat is forging rapidly ahead and the fresh wind can send the lethal hooks in strange directions.

If all goes well you may be back in harbour at 3 a.m., and you'll be free to sleep till six o'clock when you've got to be on your feet again.

You see, you've got these lines to haul and so you must be looking for your little flag buoy at the crack of dawn.

Hauling the lines is a slow business at best, but if you're lucky you may be home for dinner, probably a late one, then you'll land your few boxes of fish for the 4 p.m. sale.

That would be your day's work done but the lines have to be redd* in preparation for tonight and that'll take you an hour or two. And possibly you've forgotten that you have another bucket of limpets to get, and haven't you to meet the train tonight to collect your lug?

When do you sleep? Mostly on your feet because you and your bed are strangers. If you are to work 100 hours in a six-day week you can only sleep in snatches, but when codlings are fetching as high as twelve shillings a box you've got to keep going.

Thus the 1930s, when a pound could purchase such a lot — if only you could get the pound.

My most vivid memory of those days is the time when I was landed on the pier in the dark of a winter morning while my shipmates went back to sea to haul the lines.

I had a hook embedded so deeply in my finger that neither my mates nor I could get it out. Most of the hook was out of sight in my flesh so I had to see the doctor.

Not wishing to disturb the good man so early, I waited till daylight before ringing the bell at the surgery in Queen Street.

Dr Taylor, in his dressing gown, admitted me, ushered me into his consulting room, then had a look at my finger.

'I think we'll manage to sort that,' says he, 'but I'll need to get the lassie to hud yer han'.'

'Och!' says I, I'm nae seekin' a quine here ava!'

But he paid no heed and summoned the maid from the lobby where she was busy with her brush. No hoovers in those days!

He showed the girl how to keep my finger rigid by using her own forefinger and thumb. Of course she could turn her head away if she wanted! Then he set to work with a scalpel and laid my finger open to the bone before he could remove the hook. O boys! It was sair!

'Ye can let go noo.' he said to the girl, who bolted like a flash.

Then to me he said: 'Ye'd better tak' the heuk hame wi' ye. It's ower big for my kind o' fishin'.'

* put right

Then the kindly man took me home. That was the first time I had a hurl in a car.

Good old days? Ye must be jokin'!'

The Bridge

For the first time in his eight years, he had caught the biggest tiddler. A beezer it was. Even Mike — the tiddler champ — grudgingly admitted its superiority.

'But maybe it's the jam jar that makes it look so big,' Mike qualified.

'Some kinds of glass makes things look bigger.'

It *wasn't* the glass that made it look bigger. He had urged Mike to look inside the jam jar. And there swam surely — the king of tiddlers.

'*I* don't reckon it much.'

Anxious to keep on Mike's side, Tich McCabe peered into the jar. 'And Mungrel doesn't reckon it much either. Do you, Mung?'

Mungrel, who never spoke until somebody else put words into his mouth, agreed with Tich, 'S'right. I don't reckon it much neither.'

'Could easy not be a tiddler at *all*!'

Dave Lomax shouted from his perch on the branch of the tree. 'Could just be a trout. A wee trout!'

'— Could be...' Mung echoed; for although he had never set eyes on a trout he was in agreement with the others to 'disqualify' the tiddler.

'Let's *go* men!' Mike commanded. Suddenly tiring of the discussion.

'Scarper! First to reach the chain bridge is the *greatest*!'

'You're not some kind of wee trout.' He protested. Running to catch up with them.

'You're *not*...'

He stopped running to peer into his jam jar to reassure its occupant.

'You're a *tiddler*. And you're the *biggest* tiddler we've catched the day.'

'*Hold* it, men!' he shouted to the others. 'Wait for me.'

The authority in his voice surprised himself. Usually he was content enough to lag behind the others. Tolerated by them, because he was handy for doing all the things they didn't like to do themselves. Like swiping his big brother's fag ends. And ringing the bell of the school caretaker's door. Or handing over his pocket money to 'make up the odds' for a bottle of 'juice'. But *today* he was one of them. He had caught the biggest tiddler.

It was when he caught up with them at the bridge that his newly found feeling of triumph began to desert him.

'OK tiddler champ,' Mike said.

'Gi' us the jar. We'll guard the tiddler.'

'Yeah. Give,' Mung echoed.

'It's *your* turn to span the bridge,' Dave said.

He grasped his jar firmly against his anorak. He didn't need anybody to guard his tiddler. He didn't want to span the bridge either! *Nobody* spanned the bridge until they were *ten* at least! The others had never before expected *him* to span the bridge. He had always raced across it — the safe way — keeping guard over all the tiddlers while the others spanned it.

'*Your* turn,' Mike was insisting. '*We* have spanned it. Hundreds of times.'

'*Thousands* of times!' Dave amended.

'Even *Mungrel* spans it,' Tich reminded him. 'Don't you Mung? And Mungrel's even tichier than me!'

'Mungrel's *eleven*,' he pointed out. 'I'm not even nine yet.'

'Only...Mungrel's not *chicken*!' Mike said, 'Are you, Mung? You're not chicken.'

'I'm not chicken *neither*!' he protested. 'I'm *not* chicken.'

'OK OK!' they said. Beginning to close in on him. 'OK! So you're not chicken!...*Prove* it...just prove it...that's *all*! Span the bridge and *prove* it.'

He knew how to span the bridge all right. Sometimes — sometimes kidding on that he was only 'mucking around' he practised a little. Spanning the part of the bridge that stood above the footpath. Knowing that even if he fell he would still be safe — safe as he felt *now*. Knowing that the ground was under his dangling legs.

Left hand over right — left over right — all his fear seemed to have gone into his hands. All his mind's urgings could scarcely get them to keep their grip of the girder.

Left over right — left over —

The river's bank was beneath him now. Dark pools flowed under the bank, he remembered. Pools where the tiddlers often hid — the *biggest* tiddlers. Sometimes he had caught them just sitting bent forward on the bank. Holding his jam jar between his legs. His bare feet scarcely touching the water. He'd felt afraid then, too. A *different* kind of fear. Not for himself. Just of things which his eyes couldn't see. But which his hands could feel. Things that brushed against them...Grasping and slimy.

He would never have been surprised, if, when he brought up his jam jar to examine its contents, he discovered neither tiddler nor tadpole inside it, but some strange creature, for which nobody had yet found a name.

Left over — right — left —

The shallows were beneath him now. Looking, even from this height, as safe as they had always looked. His *feet* had always told him how safe the shallows were. A safety — perfect in itself — because it was intensified by surrounding danger.

You could stand, he remembered, with one foot in the shallows, your toes curling round the small stones. While your other foot sank into the sand — down and down...

Left over right, left over right — over —

He had a feeling that his body would fall away from his arms and hands long before he reached the end of the bridge.

Left over right — over.

It might be easier that way. Easier just to drop down into the water. And leave his hands and arms clinging to the bridge. All by themselves.

Left over — right.

He was at the middle of the river now. That part of it which they said had no bottom. That could be *true*, he realised. Remembering how, when they skimmed their stones across the water, into the middle, the stones would disappear. But you could never hear them *sound* against the depths into which they fell.

Mike had once said that, though the water *looked* as quiet as anything — far down, where you couldn't see, it just kept whirling round and round, waiting to suck anybody at all down inside it...

— over — right — left —

He wouldn't look down again. He wouldn't look down *once*. He would count up to fifty. The way he always counted to himself — when bad things were about to happen.

One two three four five six...Better to count in tens, he wouldn't lose his 'place' so easily that way—

One...two...three...four...five...six seven eight.

He thought he could hear the voice of the others. He must be past the middle of the bridge, now — the water beneath him was still black but he could see shapes within it.

The voices were coming nearer. He knew that they were *real*.

'CHARGE! MEN! Mike was shouting. 'CHARGE!'

He could hear them reeshling up the river bank. And their feet clanking along the footpath. They were running away...Ever since he could remember the days had ended with them all running away. Only *this* time, his tiddler would be with them too. And Mike would boast that *he* had catched it. He had almost forgotten about the tiddler. And it no longer seemed to matter.

Green and safe, the bank lay below him. He could jump down now. But he wouldn't. Not yet! It was only tiches like Mungrel, that leapt down from the girder, the moment they saw the bank beneath them.

Mike never did that. He could see, clear as anything in his mind's eye, how Mike always finished spanning the bridge, one hand clinging to the girder, the other gesturing, high, for a clear runway for himself before swooping down to earth again with cries of triumph!

'*Bat* Man! *Bat* Man!'

A Scottish Lion Rampant

My name is John Singh and I am twenty years of age. My father, who served in the British Army during World War Two, came to this country after the partition of the Punjab in 1947. Like many Sikhs and Hindus who found themselves in the Muslim part he was obliged to flee and spent many months in a refugee camp. My father's name is Rangit but in Glasgow he is known as John. When he came to Glasgow in 1951 he got a job as a driver in Glasgow Transport and after saving for some years went into the taxi business with his brother. They double-shifted their first vehicle until they could afford a second one.

Like all Sikh boys I am named 'Singh' or Lion after one of the ten gurus of our religion. We grow our hair long, do not shave, wear the turban and do not indulge in alcohol or tobacco. All Sikhs should adhere to these customs but some become 'tapit' or lapsed Sikhs. My father has ceased to be orthodox although like most fathers he brings up his family in the strict traditions of our race. Although he has ceased to wear the turban he is still at heart a Sikh. Anyway due to the industry of my father and uncle the family has prospered materially and we own our house even though it is a very old property.

When I was five years old I was sent to school — not a very pleasant experience if you can't speak the language and wear outlandish clothes. We spoke Bengali at home and it was not an advantage to be the only non-European in the class. The other children regarded me as a strange curiosity and my knot of hair became the target for the more aggressive. My father had told me not to fight unless I was first of all attacked and that I should bear any insults without showing anger or resentment. As time went by things gradually improved; I was good at sport, popular with my teachers and quickly picked up the language. Soon I could speak better English than my parents and thought myself superior to my old grandfather who could not speak a word of the new tongue.

When I was twelve I went to secondary school and although we learned new subjects like Algebra and Physics I had to travel further to get there. The school was in Pollokshields, a suburb of Glasgow, once respectable middle class terraced mansions but now mainly an immigrant area. Most of the children were non-European; many Pakistanis, some Sikhs and Hindus and a few Chinese from Hong Kong. The Scottish boys were always the worst behaved in the class room and the biggest bullies outside of it. Most of them had no interest in their lessons, no respect for teachers or parents and most of the time they smoked, gambled or fought in the playground or played truant for weeks on end. Neither myself nor their teachers missed them.

One day a boy two years older than myself struck me because during a game of football in the yard I had taken the ball from him. 'That was a foul, you b.... b......,' he screamed lashing out at me. I lost control of myself and hit him on the chin. He went down in a heap and it was not until he had been carried into the sick-room that he started coming round. The headmaster was very angry and I thought I was certain to be expelled but after several witnesses had stated that I had not started the trouble he calmed down and let me off with a severe reprimand. Should any trouble ever occur I was not to retaliate because I did not know my own strength. If I was in such a situation I was to see him first and not take vengeance into my own hands.

It was around this period that I passed through the Sikh process of initiation. At adolescence boys assume the name Singh and adopt the five Ks. From henceforth their hair and beard remain uncut, they wear their hair coiled under the turban (*Kesh*). At least twice a day they comb their hair with the *Kangha* or comb. On their right wrist they wear a steel bangle (*Kara*) and on ceremonial occasions they dress in the *Katchacha* or soldier's shorts and carry the *Kirpan* or short sabre. Daily they will honour their religion by reading from the Adi Granth or original scriptures, and attend, where possible the Sikh temple or *Gurdwara*. Many Sikhs are vegetarians but increasing numbers of them have ceased to be so through living in Britain and other countries.

When I was sixteen I sat my SCE 'O' Grade examinations and I thought I had performed reasonably well. During the vacation I considered whether to return for further schooling or to take up employment. I wanted to be an asset to my family and there was something grown-up in going out to earn a living. By British standards there are rather a lot of people in our household; there are my parents, my father's parents, my uncle, my three younger brothers and two older sisters. Only my father and uncle are earning money and there are a lot of expenses. My mind was made up, I would take a job.

A firm advertised in the newspaper for apprentice television engineers. Just the thing I thought. I applied and was given an interview. The man I saw was not at all friendly. I could sense that right away. I felt somehow a deeply held hostility. It was not that he said anything but there seemed something sardonic in his cosmetic smile and unnatural politeness. Perhaps it was my turban which upset him because all the time he seemed unable to take his eyes off it. At any rate my worst fears were confirmed because a week later I was informed that I had been unsuccessful in my application. All that long summer was an encore of the same story, 'Sorry the job has been filled.' By this time I had received my examination results. I had passed them all with distinction but there were still no employment prospects and I was reconciled to returning to school. There I would be at least the equal of the other immigrants.

My father, who had been following the course of events got in touch with a friend who was a member of the Race Relations Council. He advised me to apply to the Glasgow District Council whose Housing Department took on young people for apprenticeships. I did as he suggested and soon got an interview. This time it was a woman who interviewed me and she was both friendly and helpful. She asked me why I wanted the job and I explained my family's position. Then there followed an English and Arithmetic test and an aptitude test which involved fitting shapes into a board. I must have created a favourable impression because within a fortnight I started as an electrical apprentice.

This was my first experience of employment and like all beginners I had to go through the initiation of being sent on fool's errands for tins of tartan paint and left-handed screwdrivers. The squad, or at least some semi-literate member, called me Sabu after some character in Kipling's *Jungle Book*, not that they had read it but they might have seen the film. There was nothing nasty or personal in the banter; a Chinese would have been called Charlie Chan or a negro Rastus. It was a form of racialism but one in which the participants did not realize their involvement. To them it was natural that someone with a different pigmentation of the skin should be made aware of it, just like a one-legged man or a hunchback.

I had the job of making the tea, getting the rolls, sandwiches and cigarettes and carrying the tools and equipment. I had this dubious honour not because I was a Sikh but because I was the most recent apprentice and by custom and tradition these were the tasks he performed, perhaps as a means of dispelling any false pride and inducing humility.

As time went on I ceased to be the junior apprentice and got more important work to do; cutting the conduit into the appropriate lengths; screwing the ends for couplings and bending it into the required shapes; making channels through joists and walls for electrical points. It was hard work and often I would return home in the evenings totally exhausted and covered from turban to toe in brick dust and my muscles aching.

I remember one Friday afternoon we broke up early for the Christmas and New Year holidays. It was the custom to go to the public house for a drink. As a Sikh I did not have much to celebrate but I went along and although I did not touch alcohol I took an orangeade. While I was in the toilet some joker must have spiced my drink because shortly afterwards I began to feel uncomfortably warm and uncertain in my movements. Somewhat erratically I made my way home using the lampposts more for support than illumination, singing the warrior songs that I had learned from my grandfather. My father was very angry at my condition and wanted to make a complaint but once I had sobered up I explained that it was an old Scottish custom and that no harm had been intended. Nevertheless I was much more circumspect on future festive occasions.

As part of my apprenticeship training I attended a Further Education college on a one day per week basis. Here we were preparing for the City and Guilds Examination in our trade. Apart from the practical training in the workshop we were taught the theoretical basis of all things electrical and mechanical pertaining to our craft. We also spent time in what was termed, 'English and General Studies' where we were taught how to communicate both orally and on paper and how to express our thoughts with clarity for examination purposes. At least this was the intention although many students preferred to be happy in their ignorance; they considered themselves electricians and not students of language or linguistics.

One day the teacher asked if any of us from the commonwealth (he had a more delicate way of putting things than some others I have met) might be interested in writing an account of our experiences at work and school and how it felt to grow up in what might be considered to be an alien culture. I had up till that point never thought of doing so although I did have, when I came to think about it, definite opinions on the subject.

That is how I came to be writing this article. Other Asian students in the class started off with the same idea but all of them lost interest and dropped out. While the class has continued with their current affairs, memoranda and reports, I have struggled to separate the important from the trivial aspects of life and attempt to impose some coherence upon seemingly random occurrences. I cannot type but the teacher has promised to do this for me while at the same time correcting any spelling or grammatical irregularities but retaining the substance and spirit of my story.

The Scottish way of life is very different from our own. Most people are neither ambitious nor particularly hard-working. Unlike most Sikhs they have no desire to improve themselves. If for example they become unemployed they blame their bad luck or the government; they sign on at the Labour Exchange and complain they don't have enough money. My work mates, who are good fellows, work hard all week and then give most of their money, not to their families, but to the bookies and publicans. They are always, despite being in employment, short of money. It is unusual for the vast majority of them to own their own home or to ever consider setting up their own businesses. When their parents get old they are left to fend for themselves or failing that put into geriatric homes mainly because their sons and daughters are too busy leading their own trivial lives. Most people also have little concern for their children; both parents often work leaving the children to bring themselves up. Parents do not always show their children a good example and if they smoke and swear in the home it is not surprising that their children do likewise. It is not surprising that there is so much crime and juvenile delinquency because in this country what masquerades as 'freedom' is really a definite lack of concern and social control. People here do not realize how lucky they are in having a Welfare State and adequate pensions and allowances. Certainly there is recession and unemployment but there is nothing like the poverty that is found in Bombay or Calcutta. There is great political freedom and a lack of censorship yet most people are apathetic and take everything for granted. They envy the immigrants who come here poor, make a success of their lives and end up owning businesses. Scottish people seem preoccupied with the trivia of life; horse-racing, football pools, bingo and television.

My Scottish friends marry yet they don't seem to take it very seriously because they prefer to spend much of their leisure time in the pub with their mates. Before they marry they knock around with a variety of girls and go to discos and get into fights. In my opinion this seems such an awful waste of time and effort.

As for myself I shall take my father's advice about marriage. In this society people claim to marry for love then spend the rest of their lives getting over their mistakes. I do not like the British customs such as divorce because apart from it being an admission of failure, it is very upsetting for the couple concerned and disastrous for the children.

We Sikhs believe in arranged marriages. Contrary to general belief they are not forced marriages because the couple meet several times to see if they are compatible — so there is no coercion. The family merely arrange that two young people who are apparently suitable should meet because by our custom it is not only the couple who are concerned but their families also. My father says that in a few years I shall probably marry a second cousin who is two years younger than me and who lives in Birmingham. With us it is important that both families live in harmony, that the husband is caring and hardworking and that the wife is prudent and obeys.

I am often asked by British people what it feels like to grow up in a strange country whose customs are vastly different from those of India. I can only say that I know India only through the eyes of my parents and grandparents. As I have grown up here I am quite happy here, probably happier than I would be as a farmer in the Punjab which would be even more of a foreign land to me. What upsets me, although this is becoming less frequent, is when strangers appear to take a dislike to me because I am a Sikh and wear a turban and not for any harm I have done them. Once when I was working with a tradesman a woman who wanted a repair done refused to let me into her house. My mate was all for refusing to do the job but I told him not to make a fuss so he completed the job while I sat in the cafe reading the newspaper. Such incidents are few but they are sufficient to cause embarrassment. Some people are initially hostile but once they get to know you sufficiently well their prejudice diminishes and they usually end up seeing you as a human being and not some stereotype of their imagination.

In some places in Britain the National Front and the Skinheads make life difficult for the immigrant community but there is no trouble like that in Glasgow. Perhaps one of the reasons for this is that there is already enough hostility between the rival and bigoted supporters of the 'Old Firm'.

I once went to an Old Firm game between Celtic and Rangers at Ibrox Stadium. A friend of mine who is a Rangers fan invited me to go for what he termed 'a great experience'. Well it was. He gave me a Rangers scarf to wear and for once I left off my turban and put on a red-white-and-blue tammy. I was for the day an unheard of phenomenon — a Sikh Rangers supporter!

Before the game began the fans had assembled on the terracing, were drinking from cans and bottles (which is illegal but still an old Scottish custom). They were singing songs about what I thought was Delhi walls but which my friend corrected me was Derry Walls — a reference to some historic victory in Ireland — although what it had to do with a football match escaped me. By the time the teams appeared many of the supporters were already drunk and were shouting abuse at the opposing team and their supporters who were confined to the other end of the stadium. Even my friend who was normally very quiet became excited once the game had commenced. Like the rest he started cheering on his own idols and mouthing profanities at the team in green-and-white. When Rangers scored the opening goal, the fans went hysterical, hugging and kissing each other and jumping up and down. Spectators embraced like long lost lovers and hurled their tammies into the air as if to placate the gods. Finally the game ended in a draw and we made our way slowly out of the ground. Already sporadic fighting had broken out and police were arresting fans who were being bodily thrown into the 'black maria' or police wagon and taken away. In India, at the time of partition, about two million people lost their lives but I don't think the Hindus and Muslims could have hated each other to the same extent as those rival football fanatics that I saw that day. If this is sport then I will try something else. Yet strangely enough those same supporters can live and work in harmony in a normal situation.

Each week I put away part of my pay and in a year's time I hope to have saved enough for a deposit for a motorbike. By then I shall, all being well, have completed my apprenticeship and passed my City and Guilds final examination. When I have gained sufficient experience I hope someday to start my own electrical business. I could of course go into the family taxi business but I prefer electrical work to constant driving at all hours of the day and night.

There are few Sikhs in Glasgow who have become tradesmen mainly because by tradition we are warriors and farmers. But we must move with the times and accept that those of us who have left the Punjab have left that way of life and must come to terms with an industrial society. We Sikhs do not worship money or seek (no pun intended) to exploit others. We are forbidden to deal in commodities like alcohol and tobacco so we cannot become publicans or run licensed grocers or tobacconists. At the same time we have no wish to become parasites upon the the host country so we must take up occupations which are honourable and confer dignity. Young Sikhs will in increasing numbers take up trades and professions. At the same time although our ways of earning a living are changing we must, as Sikhs, retain the fundamentals of our philosophy, which are to be aware of the dangers of materialism and of pursuing false ideals. It is difficult having to live in the world, having to make a living within a society and yet avoid being corrupted by it. We must not be ensnared by Maya or the world of illusion. For us God is one and not as the Hindus believe many. God is present in all things yet is unknowable. He has no attributes yet is everything. We lead this life and continue through the transmigration of souls to seek perfection. When we obtain perfection through *Khalsa* or the brotherhood of the pure our earthly life ceases and we meet God. God can only be found indirectly through the gurus who lead us from darkness into enlightenment. Through the example of the Guru Nanak, the founder of our religion, and the other gurus we learn to tread the path of righteousness.

We Sikhs are a proud people. We do not beg from others. We give generously to charity not only to our own people but to others as well. We have no priesthood and no saints so we must make our own way.

As for politics, most Sikhs remain uninvolved. In my father's opinion this is a mistake because ultimately politics has a bearing on our way of life and the opportunities we have open to us. In immigrant areas such as Garnethill, Govanhill or Pollokshields in Glasgow, Asians standing for election, whether they be Pakistani, Hindu or Sikh, would get the immigrant vote, especially if the immigrant community saw itself as being threatened or openly discriminated against. It is now belatedly being recognised that there are not enough Asians entering the police, the armed forces, the civil service and other professions. The time is not far off when the youth shall no longer be content in pursuing their fathers' occupations — the general store keeper and the itinerant market trader.

The immigration laws may also force us into politics. We accept the fears of Mrs Thatcher and her Conservatives that Britain might be 'swamped' by an incursion of immigrants from the sub-continent. What seems very obvious is that tightening up of the laws is aimed only at the Asians; people from Canada, Australia, New Zealand, White Rhodesians and those from the European Community and even Americans seem technically free to come and reside in Britain. This is why immigration to us seems racial rather than rational. Most people in Britain seem unaware that Britain as an imperial power annexed the Punjab by force of arms after the so-called Sikh War of 1842. It was also the British who massacred hundreds at Amritsar in 1919 when General Dyer's troops fired on the crowd.

In 1947 the British left India but the partitioning of the country led to massacre on a massive scale. Not only was India partitioned but the Punjab was split between India and Pakistan. It may not have been an intentional blunder but millions were either slaughtered or dispossessed. That was the background to many Sikhs coming to Britain; they came not by choice but because they had lost everything that they valued.

Mr Powell is a bad man in the eyes of many immigrants and British liberals, but he merely says what many British people think. We recognise his honesty although we deplore everything he stands for. He is a very intelligent man but a very warped one and although he claims to have an excellent knowledge of history a few facts seem to have escaped him.

Britain is a country in decline: in population, in industry, in morale. This decline has coincided with a period of immigration and some unthinking people blame immigrants as the cause of it all! That is why the National Front and the so-called British Movement have continued to flourish when they are simply 'front' organizations for racialism and Fascism.

I am a young man brought up in this country but with roots in the old. I see the advantages and disadvantages of both societies and how we can learn from past mistakes. I claim no originality of thought because many of the opinions expressed are composite ones gained through long discussions with my father, uncle, Sikh friends and with students and staff at college. We are what our experiences have made us and as a young man mine have been limited. As I go through life my opinions shall undoubtedly alter although there is no guarantee that experience and age bring wisdom.

I have been concerned here with explaining how it feels to be a Sikh brought up in a western society which is admittedly tolerant but indifferent to the culture of others. Many Britons might say that immigrants, because they choose to come here, should adopt the cultural pattern of the British and should seek assimilation rather than retain their native individuality. To this I can only say that in these small British Isles, the English, the Scots, the Welsh and the Irish have maintained their distinctive linguistic and cultural peculiarities while becoming loosely what is a blanket term — 'British'. Should the immigrants be shown less consideration?

Finally may I conclude by saying we did not come here with a begging bowl like paupers but each with two hands willing to work. The fact that many foreigners like Reo Stakis come over here and become millionaires says much for the freedom of British society but it also suggests that British people with the right initiative and motivation could be equally successful. We, the Sikhs, are a proud people. All Sikh males adopt the name *Singh* meaning Lion and all females the name *Kaur* meaning Princess. Throughout our lives we remain lions and princesses. In Scotland your emblem is the lion rampant, symbol of a proud and dignified people. We too are lions in our own way and live and defend ourselves likewise.

The First Men on Mercury

— We come in peace from the third planet.
Would you take us to your leader?

— Bawr stretter! Bawr. Bawr. Stretterhawl?

— This is a little plastic model
of the solar system, with working parts.
You are here and we are there and we
are now here with you, is this clear?

— Gawl horrop. Bawr. Abawrhannahanna!

— Where we come from is blue and white
with brown, you see we call the brown
here 'land', the blue is 'sea', and the white
is 'clouds' over land and sea, we live
on the surface of the brown land,
all round is sea and clouds. We are 'men'.
Men come —

— Glawp men! Gawrbenner menko. Menhawl?

— Men come in peace from the third planet
which we call 'earth'. We are earthmen.
Take us earthmen to your leader.

— Thmen? Thmen? Bawr. Bawrhossop.
Yuleeda tan hanna. Harrabost yuleeda.

— I am the yuleeda. You see my hands,
we carry no benner, we come in peace.
The spaceways are all stretterhawn.

— Glawn peacemen all horrabhanna tantko!
Tan come at'mstrossop. Glawp yuleeda!

— Atoms are peacegawl in our harraban.

Menbat worrabost from tan hannahanna.

— You men we know bawrhossoptant. Bawr.
We know yuleeda. Go strawg backspetter quick.

— We cantantabawr, tantingko backspetter now!

— Banghapper now! Yes, third planet back.
Yuleeda will go back blue, white, brown
nowhanna! There is no more talk.

— Gawl han fasthapper?

— No. You must go back to your planet.
Go back in peace, take what you have gained
but quickly.

— Stretterworra gawl, gawl...

— Of course, but nothing is ever the same,
now is it? You'll remember Mercury.

FEARGHAS MACFHIONNLAIGH

Wer Da*

German Geordie we called him.
That or Herman.
A stern sentry at the base of the school stair,
directing us with stiff-armed militarism;
reprimanding us through a phalanx of uniform teeth.
A target of abuse by the miscreant throng
on forced march to our bell-shaped caverns
in the salt-mine-field of information.
A sturdy wee guy he was,
in a grey double-breasted suit
and shiny black out-dated shoes.
The pale-blue circumfusions of pupil-unfathomable eyes
quick beneath the thick ice discs
of his Himmleresque specs.
Emblematic gold on brow, wrist and finger.
Grey hair in craven capitulation before his combatant comb —
a skull-hugging helmet of silver.
A Latin teacher he was,
fond of the English word 'juxtapose'.
We would kill ourselves each time we heard it,
thinking we was saying 'just suppose'.
Thickos!
But with Roman proficiency he invaded
the legion-swallowing wasteland of our wolf-tribe,
constructing ordered roads of systematic stones.
(Though it's a while since their decline in the moorland of my
    mind...)
He used to turn up to school in a German bubble-car
— a grey Messerschmitt. (I tell no lie!)
Or else on an old black bike with a haversack on his back

*German: 'Who goes there?'

71

and clips on his ankles.
An eccentric.
Dead easy for us to make a fool of.
German Geordie.
   I learned only yesterday
   that he was an Austrian Jew
   who lost his entire family
   in a concentration-camp.
I've got to get shot of this job.
I can't take the derision.

SHEENA BLACKHALL

The Jam Jar

Mother called it sadistic, catching bumble bees in a jam jar. After all, they led a harmless existence; fat, fur-coated beings, bumbling from one flower to the next with their parcels of pollen tucked to their sides, like wealthy, jobless wives of city financiers, filling their days with shopping. They weren't bad tempered or excitable, did not waspishly dive-bomb your ears like territorial bees, those garden workaholics who think of life as a gigantic honey factory, and everything else as unnecessary and useless.

Wasps were perfect vipers when caught, or cornered. They'd ricochet off the sides of the jam jars like Kamikazi pilots, their yellow eyes two pinpricks of stinging malice, like showers of venomous hailstones. Enjoying their outrage, I would shake their indignation into fury, and when I tired of that diversion, drop the jar and run like Hell in the other direction, while the incensed hordes poured out, doubtless to impale the first passer-by with stinging wrath. Bees were less amiable than bumbles, but more so than wasps. When trapped, they were quite disorientated, were not obsessively vengeful, and when unleashed, generally zig-zagged off like confused, drunked seamen staggering round an unfamiliar port. By far the easiest catch was a nice, plump, dozy bumble. It would splutter with pompous surprise at first, and veer erratically, like a weighty helicopter, but soon accepted captivity as a *fait accompli* and sank stoically down into a tamed lethargy, the perfect prisoner.

The summer of 1964 was a bees' idyll. Hot and unusually sultry, the sun made a glorious siesta out of every noon for the people of Aberdeen; you wanted to waddle barefoot on the warm pavements, behind the pigeons, where the tar hissed in molten patches, or go where the fancy took you, like inquisitive seagulls — but nothing too strenuous, not in that sweltering heat. The searchlight of sunbeams glanced off granite mica, blinding you with unaccustomed brilliance. The gardens were buzzful of industry, the worker bees toppling over themselves to harvest their pollen, petals fingered by busy antennae, the sweetest roses, the nectar of daisy and buttercup, gleaned in a hum of industry.

I was sixteen, that awkward, argumentative age, when I thought I knew everything, but everything, and everyone over the age of thirty was an old fogey. Bee catching had long since ceased to intrigue; most girls of my age were stalking boys, though stalking bees was a great deal simpler, and much less troublesome. The jam jar contained your bee only as long as the game amused you. When it ceased to be interesting, the bee and yourself parted company with no hard feelings.

It was a strange contrast, to see the gardens in my street so full of insect life, and the pavements so bare of the human variety. My street was quite old fashioned, like a page torn from a Dickens novel; cobbles and graceful gas light incongruously stuck into a twentieth-century album. Normally, with its gulls, its granite, its gas lamps, and its elegant Episcopalian church, it wasn't a street to shun. That summer, however, it was as if an invisible drawbridge had been raised, keeping trade and commerce along the causeway to a minimum. And all because of a bug, so tiny it was invisible to the naked eye, a scrap of miniscule contagion, called typhoid.

I wish, in the interests of historical accuracy, I could describe the taste of this unwelcome visitor to Aberdeen, which lost the city a fortune in cancelled holidays and panic departures. Unfortunately, I cannot. It tasted of soft cardboard, as masticated corned beef generally tastes, when pulped together with tired, green lettuce. An exotic complaint such as typhoid should have samba'd into Aberdeen on a calypso-colourful banana boat, or rumba'd along the Aberdonian airways on a whiff of Bacardi. Instead, it slunk in, skulking inside a shipload of tinned corned beef, prepacked plague, courtesy of our South American cousins.

As the city simmered in subtropical heat, banner headlines, local and national, proclaimed EPIDEMIC in alarmist print. I assumed that foreign diseases would hunt down, first of all,

foreigners, then, presumably, the underfed and disadvant-
aged, neither of which category I belonged to. It came as some
surprise, therefore, to awake one morning to a dawn chorus of
the Peoples' Republic of Beeland, in full cry, pelting their tiny
bodies (or so it seemed) against my window pane.
The weather, too, had gone haywire, veering from volcanic-
ally broiling to chilling as a corpses' ceilidh. My mother, notic-
ing nothing amiss in either the weather or the behaviour of the
indigenous insect population, immediately phoned the doctor. He came at once, a brisk, no-nonsense, dapper little man,
who'd been a Jap prisoner of war. No stranger to the wiles of
typhoid, he'd mixed medicine in coconuts in the tropical camp
to counteract its effects. As he imparted this information, it
seemed as though his stethoscope was sprouting antennae, a
buzzing in my head mushroomed to atomic pro-
portions....'Delirious' the doctor remarked. 'Send for an
ambulance.'

A herring gull flapped me a welcome at the hospital, in med-
ical orderly white, then, unkindly, jabbed a needle into my
bum, and knocked me unconscious for several hours. When I
came round, the buzzing inside my head continued,
unabated. I shook my left ear, hard, over the pillow but
nothing, not even a mosquito, fell out. Everywhere I looked in
the ward, in accordance with the isolation, quarantine regulations of the city's official Fever hospital, there were glass windows, locked. Had I not known better, it uncannily resembled
a square, marmalade jam jar. For the first time, I experienced
a kind of panic, a fear of incarceration that was claustrophobic
in its intensity, an awful, confined, crushing sense of restraint.
I wanted out, and I wanted out straight away.

Other patients, well enough to walk, crawled around each
other like drugged locusts, eyes swollen with sleepless nights,
strangers forced together by disease. At nights, the moan and
sob of the sick, delirious women rose and fell in the ward like
an eery wind in a dark tunnel, the tight-locked windows yield-
ing neither the sun nor rain.

There was a girl of my own age in the ward, small-waisted,
black-haired, with huge, protruding eyes and thin, emaciated
arms, who lay, it seemed, in a bed of flowers (so profuse were

the floral tributes sent in by her loved one). She remarked of my flowerlessness, asking if I, too, had a boyfriend. I lied, and professed to have dozens, explaining away their non-arrival by the fact that they were all seamen (a fair lie, for a seaport city) and that one was half-way up the Congo in a tramp steamer, and the other was first mate of a whaler. For one ghastly moment, it occurred to me that whalers went out with Moby Dick... but the girl (though wanly pretty) was not overly bright, and accepted the lie quite readily. As the weeks passed, the lid of the hospital jam jar slid back a little, allowing the brief privilege of a convalescent walk, the nurses leading the patients shakily out on to the felt strip of grass which separated our ward from the mortuary, uncomfortably close. We resembled a convoy of daddylonglegs, easily bowled over by gusts of wind, tottering around like human scarecrows in our make-shift bedclothes. Because of the pressure on beds, we had been allocated the male diabetic ward, and took the air in men's pyjamas, held together with hospital safety pins.

Directly against hospital regulations, husbands and wives from different wards occasionally met up during exercise time. Weakened by illness, these reunions could be most affecting to witness, so the staff turned a blind eye to them, as long as the favour wasn't abused, and all were present when the doctors made their rounds. Having no husband, child, mother, father or elderly grandparent similarly incarcerated in the hospital compound, I nevertheless developed a wander-lust too; a desire for an area of peopleless quiet, seclusion; of aloneness and just-me-ness, filling a green space.

There, behind a kitchen shed, I found it, a goodish walk from the ward, facing the sea, and squatted down for five minutes luxurious solitude. The briny, bracing North Sea air was pure nectar after the stuffy disinfected stench of the ward. I closed my eyes in ecstasy, to savour it. I closed my eyes in ecstasy, and fell asleep....

When I awoke, the sky was cloudy, the wind was cold and the sun had disconcertingly removed itself. I had a sinking feeling in the pit of my stomach that I was set for trouble, like a skipper, anticipating a squall. I started to walk, fast, then faster, then, run.

They were waiting for me, lined up outside the mortuary, a swarm of angry, swearing, waspish women, shaking their fists in rage. I'd missed the doctors' rounds. For a time, there was talk of the special exercise privileges being suspended. I was 'sent to Coventry', and I daresay I deserved it. I think it was then that the full realisation of where I was finally hit me. I couldn't. I COULDN'T get out. THERE WAS ABSOLUTELY NO ESCAPE. The ward was a crucible of spite, where rivalry, gossip and pettiness simmered and spilled over, dangerously high in temperature. So I sank to the bottom of the jar, fascinated and tormented by the glass, an institutionalised bumble, lost and broken, and very, very, alone.

I swore then, that if ever I got out of that place, no one would turn a lock on me again.

'Oh they won't keep it up,' a doctor assured me. 'Being ill, and confined, imposes impossible strains on human beings. It's their illness that makes them vindictive. It'll pass, you'll see.'

He was right, of course. Two days after, a woman left the toilet, and omitted to wash her hands — the greatest sin you could commit in a fever ward. Instantly, communal attention switched from me, selecting a new victim to ostracise. In the old days, quite close to the fever hospital, the citizens of the town burned witches alive at the stake, the nonconformists, the eccentric, those who were a little odd, the outsiders who didn't fit in...I knew, then, that human nature never changes, that always, always, there will be victims and persecutors.

The longer I stayed in the ward the less I resisted captivity. I slept, I ate, I slept, and every waking minute was planned for me. Soon, it was the world beyond the glass that was unreal. It took a long, long time, to build up the strength to fly....

Going home at last, I made straight for the garden, and lay down on the grass by the flowers, to catch the tail of departing summer. A neighbour's child wandered in. From behind his back, he suddenly produced a jam jar; walking up to a blossoming lupin, he snatched a sleepy bumble from its perch. I jumped up immediately, smacking the jar from his hand. 'Spoilsport!' he cried. The bumble, oblivious of its narrow escape, buzzed lazily up to a cloud. Nothing should be kept in a jar, not even a bumble bee; nothing, nothing, and no one!

SHEENA BLACKHALL

An Approach to Writing

Sitting down to write 'The Jam Jar' I had three clear aims in mind:
   a) To write an account, in short story form, of the Aberdeen typhoid epidemic.
   b) To isolate that small section of my experience and to place it under the microscope of thought, so that I could pick out the dominant feelings of that time as they touched me.
   c) To go back and re-live in memory by an inner folding into myself and sinking back down the years, the sensation of being teenage ... an adult almost ready to emerge from childhood, but not totally.

These, then, were my initial intentions, the favourites at the runners' gate in my mind, lining up with some more thoughts at the starting post. Many ideas join the race when a writer's imagination takes off, but few last the course. The irrelevant, the weak, the unnecessary, all fall at the first hurdle. Also, as a visual writer, I must dress my thoughts in appropriate symbolic language. Some writers respond to words on an intellectual level alone—to their abstract concepts. To me that type of writing has no appeal; it is as arid as a desert with no oasis, as black as a cave with no lamp. But, when I read a piece of writing studded with imagery, then how the cave lights up! How much happier I am when I see the thoughts illuminated on the wall.

When I came to write this particular story, I held before me the memory of long weeks in the fever ward, where windows and exits were blocked,locked and barred to the inmates; I had before me the smothering sense of claustrophobia. The image of the jam jar, with its random, trapped, cross-section of the insect population, seemed to me to be the perfect symbol of containment.

I have a theory that, given a piece of paper and a pencil, a writer *must* tell his or her story, just as a river must flow downhill, and a tree must grow upwards. Whether the story is good or awful, writers — if they are true writers — are obliged to tell it in any way they can. Writing is also an excellent way of coping with life by taking out any troublesome experiences and observing them creatively; remoulding them if necessary as an expert potter guides the clay beneath his or her hands; therapeutic and always refreshing whether it is a hobby or a career.

These, then, are the 'whys' of story-writing. But what of the 'hows', the actual business of creating a story: what lies behind the technique? The 'hows' will vary enormously from writer to writer.

Symbols and images are very important to me as a writer, especially as a visual writer, with a keen interest in art. For me, image and thought are one. I write picto-stories, often enclosing word-portraits of characters.

A weaver sitting down at a colourful loom employs several strands of thread and, as a writer, five strands of influence are basic to my work. Of these the most important, the dominant influence, is always the symbol or image. A more subtle strand running through a story is that of the *surreal*, the subconscious, the shadow part of me that gives a story its slant. I like to follow a definite pattern in my story-telling with no loose threads, and this is directly connected with the old Scots tradition of the ballads, in poem and song, where to break the rhythm would be unthinkable. So, the symbol recurs again and again throughout the story to heighten the dramatic impact. The sounds of words is important to me and they *must* be in tune with the overall piece. Finally, I cannot write successfully in a noisy, busy environment. I require a quiet, meditative peace. Ninety per cent of writing is *thinking*, is setting up the loom, arranging the strands of the design, exploring possibilities, memories, fantasies: the chance elements that come together to create the finished tapestry.

On my father's side, my grandfather wrote traditional Scottish ballads and performed them in keeping with the strong oral tradition of the north east of Scotland. He came from a long line of hill farmers and singers. My maternal grandmother was a story-teller in her own right: unpublished, unschooled above the very elementary level, but the keeper of the family archives in that she could tell every story worth telling of my ancestors. She was never simply satisfied to say of a passing farm, 'That is Gellan', or 'That is Strathmore', but would always kindle my interest with an arresting opening — 'See yon lonely farm house up on the moor? Well, long ago... ' And I was hooked, listening wide-eyed and attentive for the inevitable story that would, if I was lucky, lead to *another* story, and another. But then, my grandmother was born into another age when people lived at a gentler pace, and conversation, and story-telling, and singing, oral communication, belonged to everyone and not to a select few — a rarified breed with the designation *writer* printed on their income tax forms.

What I am saying is this. Words and pictures are not mine exclusively. They are yours too. They are our shared heritage and are there for everyone to use, develop and enjoy to their fullest capacity. Nothing stunts creative writing quicker than condemnation. The first essay, as a child, that I found pleasure in writing was a set subject from a Primary Six teacher entitled, rather mundanely, 'My Pet'. So enthralled was I with the triumphal surge of words flowing from my pen that I remained in class over the playtime period filling an entire jotter with what, to me, was inspired prose. My teacher, much impressed, asked the headmaster to step into the room and look over my work. He was a tall, lean man, well respected in the school with a hint of fear in that respect. He wielded the belt to advantage, but was always regarded as fair in his dealings with the children, though stern.

He read the account of 'My Pet' in silence. I glowed inwardly, anticipating the honey of praise, and was a trifle disconcerted to receive instead a stinging blow around the ear with the withering comment, 'Are you aware you changed tense *three times* in this story? Have you *no* knowledge of grammar, girl?'

Needless to say my love of creative writing underwent a rapid reappraisal. There were no more playtime compositions and I clung with increased warmth to the safety of Art, which had the great merit of having no grammar, and thus no danger of throbbing ears! At the age of 15 I had begun to experiment with language, digging into dictionaries like an archaeologist. My English mistress, a published poet herself, wrote in my end of term report that my efforts were 'spoilt by irrelevance, and inaccurate, often absurdly wrong, use of words.'

I was 20 and at college before an English teacher said anything remotely complimentary about my creative writing. However, that English teacher was Ian S. Munro, the biographer of Lewis Grassic Gibbon, and his approach to creative writing did not include smacking you round the ear if you dropped your commas or blotted your full stops. Once I discovered that no one was going to hit me or write scathing reports on my progress I began, warily, to explore language again in my own time and in my own way, as language should be discovered: not as a formal introduction between strangers, but a gradual getting to know each other between friends. So far I haven't had cause to regret it. Language, now, is my closest ally and confidante.

Greensleeves

The ice-cream van it was with its harsh metallic jangle. She could only just recognize the tune for what it had once been. Not that coarse parody, stilted and mechanical, a tin brashness, a gaudiness of noise. But somewhere in it, all but lost, was an echo of something old. Still to be heard through all that distortion. The faintest of memories of times long past. She heard and she remembered. The tune was called Greensleeves.

She must have learned it first at school. So many years. Half a century and more. But further back than that it went, all sadness and grace, all minstrels and knights and ladies in high towers, imprisoned.

And this was a tower high enough. Twenty-two storeys of concrete and glass. Boxes on boxes, and hers right at the top. Halfway to heaven as the young man next door would say. People on the lift were always saying things like that. Whoever was nearest the door would press the buttons for everyone, repeating the numbers like a bingo-caller. Legs eleven they would say, or Lucky for some, thirteen. Like a language all of its own. For her floor it would be Top of the House, or All the twos, or sometimes Two little ducks. That was the one she liked best, though she thought a two was more like a swan. Two little swans. Swans were like the tune again, the way it used to be. Played on a flute, bending and gliding. Curve of the swan's neck, dipping to meet its own reflection. Ripples on an old pond. Swansong.

At first she'd been terrified of the lifts, the rickety way they clanked and jarred from floor to floor. But gradually she'd become used to it although sometimes the fear came back, especially if she was alone. The lift would creak and shudder its way up and up and she would feel the emptiness below her, increasing as she got further from the ground, suspended, supportless, a sheer black drop into nothing at all.

But the lifts were about the only place she ever saw her neighbours. It was like passing through a strange town and just catching glimpses of the people who lived there. Descending or rising through layer after layer, and every layer a few more lives she would never know. Sometimes people were friendly, but there were times too when nobody would utter a word, all lost in themselves. Strange the atmosphere when nobody spoke, every one staring straight ahead or looking down at their feet, anything rather than catch another's eye. Sometimes it was just surliness, like the young man from the seventeenth floor with his neat suit and his newspaper and his rolled umbrella. He never seemed to talk to anybody. Others she thought were just shy, like the little apprentice from the ninth. More often though, the people who didn't talk just seemed too tired or preoccupied to make the effort, coming home deadened and tense after a day's work. Then at other times it was all bustle, people babbling away, children chattering and crowding on with footballs or scooters or bikes. Sometimes it was nice at weekends, people coming home in the evening, glad for the moment to be free. She liked it then, everybody was happier, more relaxed, more ready to talk and joke and laugh. But the lift would gradually empty towards the top and by the time it reached her floor she was usually alone again. Sometimes at the weekends though it became a bit chaotic, and the lifts were always breaking down because too many people would crowd on at once. And there was always sure to be somebody loud and drunk. At times it could be really disgusting, with people being sick or even using the lifts as toilets. And then the poor caretaker would have to clean up the mess. Only the other day the big Highlander from the fourteenth had caught somebody at that.

A young boy it had been.

'Staunin therr bold as ye like,' he'd said, 'pishin in the coarner! So ah took um bi the scruff a the neck an rubbed is bloody nose it it! Told um if they acted lik wee dugs they'd get treated like them. Dirty wee tikes!'

She'd thought that was a bit cruel and maybe a bit coarse as well, but she couldn't help laughing with the others. The

Highlander was an ex-policeman. A fine big man she thought. She could imagine how angry he must have been at the boy. The lifts were always marked with their slogans, strange symbols, their own names, the names of their gangs. The Highlander called them cave-paintings, said the boys were animals. The slogans, scraped and painted, were spreading to other parts of the building. Even downstairs, right outside the caretaker's office, where the sign said No Ball Games No Loitering, the walls had been daubed and sprayed. But at least it wasn't as bad yet as other blocks she'd heard about. According to the papers and the television, some of them were already on the way to being slums. She'd even heard people on the bus talking about these blocks here as if they were slums. Two women it had been, sitting just behind her.

'And I mean,' one had said, 'you should see some of the children, running about wild. Like wee savages. And as for some of the language!'

'Mind you,' the other had said, 'you can't really expect anything else. They're all just shifted out from Partick and Govan, and all these dirty old tenements are just falling to bits. It's not as if they've ever known anything better.'

She thought of her own home over in Ibrox. Near enough to Govan, but nowhere near to being slums. A decent red sandstone block, clean and solid and old, with tiles in the close and polished wooden banister-rails and stained-glass windows on every landing. The one on her landing had the head and shoulders of a young knight or squire, panels of twining flowers on either side, yellow, red and green, and up above patterns of red and blue squares. The whole close was always quiet, and the daylight that filtered through the windows was tinted and softened by the coloured glass. The back court itself, when seen through windows like these, was a bright painted garden, a landscape tinted and framed.

That was how she liked to remember it. Like another age. The stillness of a summer afternoon. It had been a fine building. Even the storm of a few years back hadn't damaged it at all. But the area was due for redevelopment, so down it had to come to make way for a supermarket or a wider road or another block of flats.

One afternoon, a month or two after she'd moved, she'd been back in Ibrox to go to a jumble-sale. When she'd passed by they were pulling the building down, grey rain drizzling on the wasteground where half a street had been. She'd stood and watched as the workmen heaped spars of wood on a bonfire and bulldozers and lorries cleared away the rubble. The ground had shaken, a terrible rumble and crash as another wall crumbled and fell in a cloud of choking yellow dust. The dust cleared and the building stood, laid open like a doll's house, and there where the second storey had been was one wall of her home, old layers of wallpaper peeling, tattered edges flapping in the wind. And where the fireplace had been was the back of a blackened hearth and a broken chimney and a line of soot left by every fire she'd ever burned. And the whole ruin flickered and wavered in the haze of the workmen's bonfire, crackling and hissing and spitting in the rain.

A fire was something she missed here, the glow of it in the evening, the stories she could read into the dance of the flames. Now she had the television to look at and the under-floor heating to keep her warm. It was good though on winter mornings, not to have to rake out ashes and footer with paper and sticks and fumble with matches when her fingers were so numb and cold. But she did like a fire to sit by. Still, the heating was good for her plants. She had four, in pots along the win-dowsill — geranium, some ivy, morning-glory and a small spiky cactus — and all of them were thriving. In the geranium pot there were even some weeds flourishing, tiny blue flowers and a thick tangle of leaves. She hadn't had the heart to cut them, she'd just let them grow like the others. She thought the seeds must have come drifting in through the window, and by some miracle had settled on this few inches of soil, and that was reason enough to let them live.

She stirred herself to get up and go into the kitchen. She filled a small milk-jug from the tap and, carrying it through, watered each plant in turn and stood looking at them, con-tent, the shapes of flower, stem and leaf, just so, against the evening sky. Somehow she always thought it added some-thing to the landscape, to see it framed by such graceful

shapes, to see them silhouetted against it. And the view from
these windows was one thing she was grateful for, some con-
solation for all the rest. She would never tire of looking at it,
the houses so far below stretching away on all sides. And there
were the hills of Renfrewshire, away to the left across the river,
and there to the right the hills of Dumbarton and Kilpatrick. And
from here you could still see it was a valley, the river winding
down through it, now hidden behind buildings, now glinting
and flowing through shipyards or open ground, past tenements
and factories, on through patches of green, traces of how it had
been before, when all of it was farmland, or wild, like the hills.

Sometimes she was amazed at how much there was to see,
how much there was going on all at once. She would be looking down at the blue train clattering out towards Balloch, she
would follow it out till her eye was caught by faraway streams
of traffic, then a gull would flap and glide up past the window,
up into the open sky, the sky she could look at forever. All day
long she could watch it flowing, watch the weather turn and
change, lose herself in the colours and shapes. Now the
clouds would come bulking and threatening up the valley,
bringing the rain up steadily from the west, now the sun
would light on a patch of hillside away across the river, now
the clouds would drift on and the whole sky would clear.

The sun had gone now, down behind the hills, and all the
street lamps were lit. By the light of the lamps nearest, a few
small boys were still playing football, directly below in the
playing area, a fenced-in rectangle of red ash. They weren't
allowed to play on the grass but they didn't usually pay much
heed and the caretaker was forever chasing them. They were
always up to something. If it wasn't trampling the grass it was
barging about the drying-area upstairs, or running and yelling
in the entrance-hall, or playing in the lifts. One of their favourite tricks was to press all the buttons so that the next one to use
the lift would stop at every floor. That was one that seemed to
annoy everybody, and the boys were always in danger of
being thumped for it.

A sudden rap at the letterbox made her jump. She wasn't
used to having unexpected visitors and she was flustered,
wondering who it could be, as she hurried to answer it. It was

the young man from next door.

'Hello,' he said, smiling. 'Ah wis wonderin if ye had any change fur the telly.'

'Oh, yes,' she said. 'Ah'll just away an have a look.'

She came back fishing out coins from her purse.

'Here you are,' she said. 'A two-shilling bit. Ah've actually got two but ah'll be keeping one for maself.'

'One's jist fine,' he said. 'Thanks a lot. Here yar then, ten pence.'

'Of course,' she said. 'Ah don't think ah'll ever get used to this new pence buisness.'

'Aye, it's a lot a old nonsense,' he said.

'Is there something good on the television the night?' she asked.

'Ah don't really know,' he said. 'We shove it oan at teatime an that's it oan fur the night.'

'Och well,' she said, 'it's good company.'

'It is that,' he said. 'Well, thanks again.'

'Not at all,' she said. 'Cheerio!'

'Cheerio!'

From the newspaper she saw that there was a quiz programme just finishing. Then she had a choice between a programme on current affairs, a comedy show and an old film. She decided to watch the film. She had seen it years before, but that didn't matter. She could barely remember it. It was called *Grand Hotel*. She liked the old films best. She hadn't been to the pictures in years.

She put her coin in the slot at the back of the television set. At least it would be enough to last till the end of the film. The night before, the money in the box had lasted right up to the beginning of *Late Call*. The young minister had started very earnestly. 'Tonight,' he said, 'I want to talk specially to those of you out there who are old, or lonely, or sick.' Then the money had run out. The minister had disappeared with a ping, shrunk to a white dot, nothingness, the screen suddenly black. That had given her a funny feeling. She'd thought of the minister in a studio somewhere, smiling at a camera, mouthing into a microphone, not knowing that she could no longer see or hear him. Those of you out there.

When she switched on the set, the quiz programme was coming to an end. The quiz-master was grinning. 'And so, from all of us here to all of you out there, Good luck, Goodnight, God bless and See you next week. Bye!' Then the theme music started and the credits rolled up. Glitter of prizes. Treasure. Applause from the studio audience. See you next week. All of you out there. That was what the minister had said. Out there. God bless.

The advertisements came on then. The first was for packets of powdered soup, an old-fashioned iron pot bubbling away over a blazing log fire, a scrubbed wooden table, a heap of fresh vegetables. Rain on the window, a hungry family, steam rising from the pot. 'Country Fresh,' said a deep, rich voice.

The next was for soap powder, though she always confused it with one for life insurance and another for margarine. It showed a young family, all dressed in white, leaping and running across fields in slow motion. There were shots of the sun, caught in the trees, gleaming modern buildings, an aeroplane, a white bird. The music was clean and bouncy and bright. A voice was saying 'Tomorrow's world ... Today!'

She was startled by another rap at the door. This time she was even more confused. Two visitors in one evening. The place was never usually so busy.

It was a man selling locks.

'Ah fit it fur ye as well,' he said. 'That's included in the price.'

'Well, there's nothing really wrong with the lock ah've got,' she said.

'Ah but this is a special mortise-lock,' he said. 'It turns four times. They don't make them any merr burglar-proof than these Missis.'

'Well, I don't really think ...' she said.

'Please yerself Missis,' he said. 'But ah hope ye don't mind me askin ye, d'ye jist live here yerself?'

'Well, yes...' she said.

'Aye well,' he said. 'Ye only huv tae read the papers these days tae see the kind of things that's happenin aw the time. D'ye know whit ah mean?'

'That's true,' she said. 'Maybe ah'll think about it. But ah don't know if ah can really afford it anyway.'

'That's nae bother,' he said. 'Ye kin pey it up every week.'
'Well,' she said. 'Ah'll see.'

'Fair enough Missis,' he said. 'Ah'm roon here every month.
Ah'll maybe catch ye in again sometime.'

'Right,' she said. 'Cheerio.'

'See ye,' he said.

She closed the door and heard him rapping her neighbour's
letterbox.

The film had just started and she settled down to watch it,
and as she watched she began to remember it. Not clearly, it
was too long ago for that, only a vague stirring as the story
unfolded, the hotel and all the people in it, never really knowing each other, all their lives so separate but intertwined. And
there was Greta Garbo, all sad and lovely and wanting to be
alone, and the thief she was to fall in love with. She remembered that. She remembered they wanted to run away
together and she remembered that somehow he was to die.
And there was the man with the strawberry birthmark. He just
sat and watched it go on all around him, and all he ever said
was 'Grand Hotel. People come. People go. Nothing ever happens.'

At the advertisement break, she got up to put on the kettle
for some tea. Then she heard the tune again, Greensleeves,
clangouring through the night air, and she remembered she
wanted to buy some milk from the van. In the evenings it came
round every hour. She was glad of that; there were no shops
open nearby; no Handy Stores or Indian grocers here.

She put on her coat against the chill and hurried out to the
lift. When she pressed the button, the light didn't go on. She
pressed again and again but it was no use. Only the click of the
button and a dead, heavy clanking, echoing up from far
below. Both lifts must have broken down. She pressed her face
up close to the small window, peering into the shaft. She
could hear people on the other floors banging the doors in
annoyance and raising their voices. The cables just hung,
swaying slightly in the gloom.

Sadly, she turned away. Even if she could have managed to
get down the stairs in time to catch the van, she would never
manage to get up again. Twenty-two flights was too much for

her to cope with. Luckily she had never had to try. The lifts had always been fixed in time. But tonight it was getting late. The workmen might not come till morning, and she couldn't take the chance. She was stuck here. Trapped. Twenty-two floors. All the twos. Two little ducks. Half way to Heaven. Top of the house.

She stopped and looked out the landing window. The boys had all gone and the playing area was empty. Directly opposite was the next block, lit up against the dark. She remembered then how she'd seen it once, early on a winter morning. It had been so dark and foggy that she couldn't make out the outline of the building, only the blackness and the lights. At first it had looked as if the lights were just suspended there, hanging in the air. Then she'd seen it as if the whole sky was one vast black wall, with these few lights set in it. That had given her the same empty, bottomless feeling she sometimes got on the lift.

She had left her door open and she could hear the television set blaring. The soap powder advert was on again. She recognized the music. Tomorrow's world ... Today. The light in the corridor was bright and cold.

From outside she heard again the din of the tune from the van, moving on to the next block. She was never quite sure of the first words of the song, whether it was Alas my Love or Alas my Lord. Alas my Love you do me wrong, to cast me off so discourteously.

The film would be starting again. She would go in and watch it till the end.

She closed the door and turned the key, locking herself in for the night.

JANETTE WALKINSHAW

The Time my Friend Mabel became an Anchorite

When my friend Mabel decided to be an anchorite*, her husband said it was all right by him as long as she paid the £1.37 out of the housekeeping and did she realise she would have to give up the drink. She said the second bit was right enough though that was maybe a simplified way of putting it, but she was damned if she could see where the £1.37 came in.

It turned out that Phil thought an anchorite was the same as a rechabite** and his knowledge of them was confined to his Uncle Archie who paid into it for years and got quite a lot of sickness benefit back. The family said it wasn't right since it was the alcohol made Uncle Archie sick in the first place, but even his best friend couldn't accuse him of having indulged after he joined.

Anyway, I was telling you about Mabel and how she took to religion. Phil telephoned me and when I went round I found Mabel sitting there, looking out of the window with her housework not done or anything. That's not like her so I asked if she was feeling ill, and she said no, she had decided to renounce the world, that was all. She showed me a book she'd been reading and it was all about saints and martyrs, and there were these anchorites who shut themselves up in a cell and never saw a human being ever again. I've decided that's what I've been looking for, she said. Well, I tried to be as patient with her as I could. After all, when I look at a cookery book it doesn't make me want to be a fish. There's no such thing as a twentieth century anchorite, I said. Yes there is, she said, and I'm it. I've had enough of people. I'm going to spend the rest of my life in total solitude and silence, contemplating the infinite. I said was she going into a convent and she

* a recluse or hermit
** a member of a benefit society and a teetotaller

90

said no, the spare bedroom. I thought about this for a while and then said in that case could I have her long blue dress with the silver sleeves and sequins down the front. It would need taking in at the waist but luckily she's the same height as me.

Phil wanted to send for the doctor or the social work department but I thought the doctor would prescribe tranquillisers and in my opinion she looked too tranquil already. Anyway, she's always been an independent person and she wouldn't like other people knowing her business. Poor Phil. I found she hadn't made him anything to eat, so I cooked him a nice cheese omelette which I always find very acceptable in emergencies, and he enjoyed it and felt much better. He had a bowling match on so he went away and I went up to the spare bedroom to have a serious talk with Mabel.

I mean, it was obvious what the trouble was. A lot of people want single beds by the time they reach Mabel's age and you don't have to make such a big thing about it. I told her this, through the keyhole, for she had locked the door, but she said I was mistaken, it was nothing to do with that. She just wanted nothing more to do with people. She had never liked them. The rest of her life would be spent the way she wanted it, in seclusion, and we could leave her food on the landing.

Well of course we thought she'd get over it in a day or two but she didn't. Phil thought if we didn't give her any food she would soon be hungry and come out so we tried that, but Mabel was carrying quite a bit of spare weight. She always has been very self-indulgent as this whole business shows, and it would have done her no harm to go hungry for a while. But of course the people who came to see her brought food, sometimes just an apple or a piece of cake, sometimes a boeuf en croute with asparagus, but it was enough to keep her going.

You see, word got round and people started coming. I don't know what they expected to find, or what they expected to get out of it. They sat on the landing waiting for Mabel to speak which she hardly ever did, but that didn't worry them. They would tell her their problems through the door, and sometimes she answered them and sometimes she didn't but either way they seemed quite happy.

It was nice for Phil to have company. He took the whole thing quite well, I thought. You know me, he said, Phil for philosophical. It's one of his jokes. Many a grand night we used to have at the bowling club. He was the life and soul of the party. Mabel sometimes said she wished he would get some new jokes, but then she never was one for parties, and sometimes tended to be a bit of a wet blanket after midnight, but the rest of us thought he was a laugh. We used to shout out the punchline before he came to it. It was great fun. Anyway, he was being very philosophical about Mabel's defection and I did what I could by taking his socks home to darn.

Some Americans stopped off on their way to India. They were very interesting and not at all what you would expect by looking at them, but then they are a nation of extremes, aren't they? It was a bit of a nuisance for Phil to find nothing but goat's milk and dandelion root in the fridge when he was hungry. They sat in a circle round the house chanting, a low moaning sound. It was creepy till you got used to it. In fact you missed it when they stopped. They didn't often stop for they used to moan in relays and it was only when the moon and stars were in a certain conjunction they would stop for five minutes every hour. One of them explained it to me. It was quite exhilarating. You could hear them all the way to the by-pass.

Then there was Mr Smith from down the road. To tell you the truth I don't have much patience for people who have hobbies that make them miserable, but you have to feel sorry for Mr Smith. His spare time was spent phoning up women and breathing heavily. Of course they hung up on him immediately and it made him very frustrated. He came to see Mabel and breathed at her through the keyhole for hours and hours, and went away a happy man. He said it changed his life having a woman listen to him without interruption. He had never believed it possible. He came back several times, but he's on shift work and couldn't always manage.

The People's March for Jobs made a detour to see Mabel and stayed for a while. Their leaders stood below her window to address the crowds. Personally, I think it was mean of her not to speak to them. She'd had plenty of time to contemplate the world's problems, and come up with an answer. If she had she kept it to herself.

I was interviewed by News at Ten standing under Mabel's window. I had my hair done specially and wore my green wool suit. I said it was a good thing having an anchorite in the middle of a housing scheme, for it made everybody feel more uplifted. I said the government ought to sponsor a Nun-in-Residence in every town and do away with the National Health Service. I said what's more the vandals weren't getting peace with all the people about and went somewhere else, so it was an answer to the vandalism problem as well. I was on just before the adverts and a friend took a video so I can show it to anyone who missed it.

A psychologist from the university came. He sat on the landing and asked Mabel a lot of questions about her dreams and the change of life. Of course, she didn't answer but he just sat there repeating his questions over and over again through the door. He was a persistent man, and reminded me in many ways of my late husband. We could hear Mabel humming to herself, and this made the psychologist very excited. Since she didn't answer him he started asking the people in the queue about their menstrual problems. Oh, we had some lively discussions there on the landing, I can tell you.

Poor Phil was getting a bit fed up by this time. I found him one day in the kitchen very disconsolate. The people from the women's magazine who wanted to do Mabel's life story had made an awful mess of the garden with the elevating platform they borrowed from the motorway lighting men to reach Mabel's window. Not that it did them any good for she just drew the curtains and refused to speak to them though they sat outside her window for days, but one of the wheels of the machine went over Phil's prizewinning marrow. At least it would have been prizewinning if it had survived to see the flower show. It turned out as well that all his winter jerseys were kept in the spare room and Mabel wouldn't let him in. With the weather turning a bit colder he was needing them for the bowling. That's another thing, he said. You can't get a decent game any more for everybody wanting to talk about Mabel. I comforted him for a while, and said it was my opinion that nobody could stand their own company for so long and she would soon come out.

Of course, it was inevitable that someone in higher authority would start to take an interest, so that when we saw the Royal helicopter hovering over the house I wasn't a bit surprised. I'm very sensitive to these things and I knew it was just a matter of time before Mabel received official recognition.

Well, the helicopter was hovering and looking for a place to land, and taking a while about it, for I suppose he was more used to the lawns at Buckingham Palace and the gardens round here are very small. Everybody rushed out of the house to watch him.

Except Phil. He grabbed my arm and dragged me upstairs!

But he only wanted him and me to talk to Mabel without all the people around. He pleaded with her through the door till I got fed up and I could hear the helicopter had landed, so I used my credit card to spring the lock, for it opens all doors.

And we went into the room. And she wasn't there. I was quite speechless.

Her blue dress was hanging on the wardrobe with a note pinned to it. It was addressed to Phil so I read it out to him. *Dear Phil, I can't bear it. I have gone to Novaya Zemlya. Your winter woollies are in the second drawer on the right. Mabel.*

Well, I did the only thing possible in the circumstances. I mean, what would you have done? Everybody was coming, so I pushed Phil out of the room, and closed the door at our back. We waited on the landing to give the appropriate welcome.

The rest is history.

Everybody thinks Mabel's still there. It would be a shame to disillusion them, for it makes them so happy. Besides Phil's doing awfully well with the collecting boxes, and my cream teas at £2 a time have become a legend. On the other hand, I'd quite like a holiday, but I daren't leave Mabel unattended.

Still, I'm glad she left me her blue dress. It fits me beautifully, after I took it in at the waist, of course.

Snowballs

'The minister, the Reverend Murdo Mackenzie, and his son, Kenneth, will be visiting the school tomorrow,' Mr Macrae told the boys of Standard Seven. 'I want you to be on your best behaviour.' They sat two to a seat in a room which was white with the light of the snow.

'That is all I have to say about it,' he said. He unfurled a map which he stretched across the blackboard. 'And now,' he said, 'we will do some geography.'

The following day was again a dazzle of white. Mr Macrae took a watch from his breast pocket and said, 'They will be here at eleven o'clock. It is now five to eleven and time for your interval.' They rushed out into the playground and immediately began to throw snowballs at each other. They would perhaps have a longer interval today and then Mr Macrae would blow his whistle and they would form lines and march into the room.

Torquil shook his head as he received a snowball in the face and then ran after Daial, whom he hit with a beauty. The sky was clear and blue, and the snow crisp and fresh and white.

At eleven o'clock they saw a stout sombre man clad in black climb the icy steps to the playground, a small pale boy beside him. They stopped throwing snowballs for they knew this was the minister. He halted solidly in the middle of the playground and said,

'This is my son Kenneth. I shall leave him with you for a while. I am going to see Mr Macrae.' He had a big red face and a white collar which cut into his thick red neck. Mr Macrae was waiting for him at the door and they saw him bend forward a little as he welcomed the minister into the school.

'Come on,' said Torquil to the small pale Kenneth. 'You can join in if you want.' Kenneth seemed at first not to know what to do, and he stood uncertainly in the middle of the playground, while all around him the boys whirled and shouted and threw snowballs. Then he too began to throw snowballs and after a while he was enjoying himself hugely, and his pale face glowed with colour. He got a snowball in the back of the neck but he gave another one back, though he slipped once or twice being not at all sure on his feet. He ran almost like a girl with his hands in front of him. The interval passed quickly and then just at the moment that Kenneth had received another snowball, this time on the cheek, Mr Macrae and the minister appeared at the door.

They saw the minister stride wrathfully forward after saying something to Mr Macrae and still with the same uninterrupted stride descend the icy steps, his hand in his son's hand, and disappear from view. Mr Macrae blew his whistle and they all lined up, still red and panting from their exertions. As they stood in line they saw that Mr Macrae was trembling with rage and his face was white.

His moustache bristled as he shouted,

'So you threw snowballs at the minister's son, eh? Eh? I will teach you.' While they still waited in line he went furiously back into the school-room and emerged with a belt. 'So that you will know what you are being belted for,' he said, 'you are getting it for throwing snowballs at the minister's son. You have made me a laughing stock. The minister's son is at a private school and is not used to such behaviour. You have shown yourselves to be hooligans, that's what you have done.'

'Hold out your hand,' he said to Daial, who was at the head of the line. They heard the belt whistle through the air six times. 'Now your other hand,' said Mr Macrae. Torquil waited. He was sixth in line. He knew that the belt would be very sore on such a cold winter's day. He spat on his two hands in preparation. Swish went the belt and the more he belted the more fierce became Mr Macrae's rage.

While he was waiting to be belted Torquil said to himself, 'Kenneth was enjoying the snowballing. Why are we being belted?' But he knew that the belting didn't have anything to do with Kenneth, it had to do with the minister, and perhaps not even with the minister but with Mr Macrae. Deep within himself he felt the unfairness of it: twelve of the belt for throwing snowballs, that was not right. Ahead of him he heard the Mouse whimpering quietly and saw him bending down, wringing his hands as if he were in unbearable pain. Mr Macrae now reached him and said, 'Hold out your hand, boy,' not 'Torquil' but 'boy'. He did so and the first stroke had a sting that made him wince. The second one was worse and by the time that the sixth one came he felt that his hand had been cut. He gritted his teeth as tightly as he could. 'The other hand,' said Mr Macrae, his small pale moustached face fierce and determined. The belt rose and fell, rose and fell. For one crazy moment Torquil thought of withdrawing his hand and then decided against it, even though he was as tall as Mr Macrae if it came to a struggle. Then it was all over and they were back in the classroom again.

They sat down in their seats and for a while Mr Macrae turned his back on the class, breathing heavily as if still not satiated.

Many of the boys were wringing their hands under the desks and the Mouse was still whimpering quietly.

'That's enough,' said Mr Macrae and the Mouse stopped whimpering.

The boys opened their poetry books and they read round the class.

> 'A wet sheet and a flowing sea
> a wind that follows fast
> and fills the white and rustling sail
> and bends the gallant mast
> and bends the gallant mast, my boys,
> while like the eagle free
> away the good ship flies and leaves
> old England on the lee.'

Mr Macrae beat with his ruler on the desk as if it was a metronome.

> 'There's témpest in yon hórned móon,
> and líghtning in yon clóud
> But hárk the músic máriners
> the wínd is píping lóud.'

Suddenly he seemed to have become jolly again and to have forgotten the belting and the snowballing.

'The wínd is píping lóud, my bóys,
  the líghtning fláshes frée,
while the hóllow óak our pálace ís,
  our héritage the séa,'

Torquil put his hand up.
'What is it, Campbell?'
'Please, sir, I want to go to the toilet.'
'All right then, all right then,' said Mr Macrae in the same jolly voice. Torquil left the room and went outside into the whiteness. It was snowing gently and the flakes broke like stars on his jacket and his white trousers. The toilet at the back of the school was cold and draughty, and there was no lock on the door. The water poured down the walls. He stood there for a while contemplatively peeing, his hands so raw and red that he had difficulty in unfastening and then fastening his fly.

After a while he left the toilet and went into the school again. As he was coming in the door he saw Mr Macrae standing there, while from the classroom whose door was shut he could hear the boys chanting in unison.

'O for a sóft and géntle bréeze
  I héard the fáir one crý
but gíve to mé the snóring bréeze
  and whíte waves héaving hígh.'

'And white waves heaving high,' said Mr Macrae jocularly. 'So you will throw snowballs at the minister's son.' And he made to hit Torquil on the bottom with the belt, but Torquil slid away and was hit on the head instead. For a moment a phantom fighter turned on Mr Macrae and then he was back in the classroom again, sitting in his seat. Mr Macrae was now in a good mood and shouting,
'And bends the gallant mast, my boys. Can't you see it, boys, the ship with all sails set crowding across the ocean. The storm can do nothing to her for as we are told in the poem the good ship was tight and free. What is the hollow oak, Torquil?' Torquil looked up at him out of the gathering swaying dark-

ness into which he abruptly fell, the ocean closing over him.

'Torquil,' shouted the headmaster and then there was complete darkness. Later he felt himself being set on his feet. Cold water was streaming down his face. Mr Macrae was speaking to him nervously.

'Are you all right, Torquil?'

'Yes, sir.'

'Good, good. You can go home then. Did you hear me? You may go home. Tell your father I shall be along later.' Torquil stood on the floor, no longer swaying.

'Fine, fine,' said Mr Macrae, 'it was an accident, you understand.'

Torquil left the classroom and walked across the playground and down the steps and then turned left to go home. The snow was still falling, very lightly, on his jacket. Soon it would be Christmas, he thought.

When he went in the door his mother looked up. Her hands were white with flour.

'I'm not going back,' said Torquil.

'What did you say?'

'I'm not going back. Ever,' said Torquil.

'I am going for your father,' said his mother, and she went into the byre where her husband was busy with harness.

'Torquil has come home,' she said, 'and says he's not going back to school again.'

Her husband raised his grave pale bearded face and said, 'I will see him. Tell him to come in here.'

She went back into the house and told Torquil.

'Your father wants to see you in the byre.'

Torquil went into the byre where his father was waiting. The smell of leather calmed him: he would like to learn how to plough. Next spring he would ask his father to let him.

'What is all this?' said his father. 'Sit on that chest.' Torquil sat down.

'Well, then,' said his father.

Torquil told him his story. He tried to tell his father that the worst part of it was not the belting but the difference between him and the minister's son, but he couldn't put into words what he felt. He put his raw hands under his bottom as he sat on the chest. His father didn't say anything for a long while and then he said,

'Mr Macrae is a good man. He is a good teacher.'

'Yes, father,' said Torquil.

'The one before him was too slack.' Then he stopped. 'I will think about it.' Then, 'Mr Macrae is a good Navigation teacher,' he added as if this was as important. 'Go inside now.'

At half past four Torquil saw Mr Macrae heading for the house on his bicycle, a small figure on which the snow was falling. Through the window, itself almost covered with snow, he saw him approaching and then his father going to meet him. He couldn't hear what Mr Macrae was saying but saw that he was gesticulating. His father stared at the ground and then shook his head. He seemed to be much calmer than Mr Macrae who was like a wasp humming about a bull. Then after Mr Macrae had talked a great deal, Torquil saw him get on his bike and ride away. After he had gone his father sent for him.

'You are not going back to school,' he said. 'You will work with me on the croft. We will say no more about it.'

Torquil saw that his mother was about to say something but his father looked at her and she bent her head to the plate again.

That spring Torquil was allowed to help his father with the ploughing, which was harder than he had thought. The plough refused to go in a straight line, the patient horse tugged and tugged. Seagulls flew about the sparse ground, and a fresh wind was in his nostrils. Sometimes as he walked along he could hear a voice in his head saying

'And bends the gallant mast, my boys,
while like the eagle free
away the good ship flies and leaves
old England on the lee.'

The black earth turned and the blades were hit by stones. He felt as if he was captain of a ship, his jersey billowing in the breeze.

'You'll come on fine,' said his father and then to his mother that night. 'He's coming along fine.'

When he was eighteen years old, because there was no

employment, he decided to emigrate to Canada. He stood on the pier, his father and mother beside him. The ship's sails swelled in the breeze.

'You will be all right,' said his father. 'You have a good grounding in Navigation. Mr Macrae saw to that.'

'Yes,' Torquil agreed.

He went on board the ship after kissing his mother and shaking his father by the hand. As the ship sailed away from the pier he saw them standing there with a lot of other people who were seeing relatives off. The sails swelled and soon they were far from shore and the island was a long line of green with lights twinkling here and there. Then it could not be seen at all.

He had a hard time of it in Canada for it was during the thirties that he emigrated. Sometimes he slept in doss-houses, sometimes he worked on the railway tracks. At nights he and the other boys from Scotland kept themselves warm by dancing the Highland Fling. His underclothes were in rags and one morning in spring after washing them in a stream he threw them away. Eventually he reached Vancouver and there got a job as a Fire Officer. He trained hoses on charred bodies in burnt rooms.

One night at a ceilidh in another islander's house he had an argument with him about the Garden of Eden.

'It wasn't an apple that was mentioned,' he said. 'It was just any fruit. I'll show you.' And, after asking him to get his Bible, they both studied it. It didn't mention an apple at all. It simply said the Tree of Good and Evil.

'You know your Bible sure enough,' said the islander whose name was Smith, and who was lame because of an accident on the grain elevators.

Torquil didn't say anything.

'The funny thing is that I never see you in church,' said Smith.

'You will never see me in church,' said Torquil.

But he didn't say why not. It seemed to him strange that he felt no anger towards Macrae, whom he still regarded as having been a good teacher, especially of Navigation. Sometimes when it was snowing gently he would see the belt descending, he could hear the words of that poem which he had never forgotten and he could see the thick neck and face of the minister.

'No,' he repeated, 'You'll never see me there.'

TOM LEONARD

Four of the Belt

Jenkins, all too clearly it is time
for some ritual physical humiliation;
and if you cry, boy, you will prove
what I suspect — you are not a man.

As they say, Jenkins, this hurts me
more than it hurts you. But I show you
I am a man, by doing this, to you.

When *you* are a man, Jenkins, you may hear
that physical humiliation and ritual
are concerned with strange adult matters
— like rape, or masochistic fantasies.

You will not accept such stories.
Rather, you will recall with pride,
perhaps even affection, that day when I,
Mr Johnstone, summoned you before me,
and gave you four of the belt

like this. And this. And this. And this.

Mossy

Miss Stone had her departure from the school nicely planned. At one minute to four, she lined them up. Ten seconds before the bell, she sent a boy to spit his bubblegum into the wastepaper basket. Then she dismissed her very last class. They walked demurely down the corridor, then roared off round the corner. Miss Stone, a complacent smile on her neat, blunt features, picked up her coat and brief-case, and nipped across the corridor into the Detention Room.

There, studying the Detention Book, was Mr Downie, youngest and newest teacher on the Staff.

'Mr Downie,' she said. He jumped, and Miss Stone smiled grimly.

'As you know, I retired from teaching at 4 pm today.'

Mr Downie appeared to be at a loss.

'You're not having a party, then?' he said.

'Certainly not. The Rector being unctuous, concealing how much we dislike each other; other members of the staff who were once pupils here — an incestuous situation, don't you think? — would recall how I once belted them. No thank you. Now, I would like —'

'I was once a pupil here,' said Mr Downie, 'and you belted me.'

'Indeed? I didn't realize. But then I have never involved myself with my former pupils.'

'You weren't much interested in your current ones.'

Miss Stone's eyebrows rose. Mr Downie was evidently made bold by the prospect of her departure.

'I'm sorry I haven't time for a stroll down Memory Lane,' she said. 'I have a favour to ask of you. If you would be so good —'

'If I would be so good! Always the soul of courtesy, you

were. Till there was trouble. Then it was a hand like a breeze block rearranging your brains. I think you impaired my hearing permanently.'

'Then I must speak louder. I would like to take your Detention, this evening.'

There was a bewildered silence.

'You mean do my Friday night Detention for me?'

Miss Stone sighed and waited.

'Why? I mean...I've been lumbered with it all term, and nobody's ever agreed to swop with me. Who would? Dammit, it's Friday, isn't it? And,' he went on, marvelling, 'you wouldn't be swopping if you're leaving today —'

'Mr Downie,' said Miss Stone, losing patience, 'I wish to avoid certain contingencies. Such as the medium sweet sherry, now awaiting me as a surprise in the Ladies' Staff Room. Another is having my hand shaken warmly by colleagues whom I have detested for years. In short, I want to hide until everybody has gone home. Will you assist me?'

Mr Downie grinned broadly.

'You should have been called Stony, not Mossy,' he said 'You always had your own way, didn't you? I wish I had the knack. The kids terrify me.' He glanced down at the Detention Book and shuddered. 'Especially the girls. One thing I always appreciate, though — your sardonic sense of humour.' His own humour overcame him. 'OK, sister,' he said, 'you talked me into it.'

Miss Stone waited, looking down her nose.

'Yes, yes, thank you. I'll do it,' he said, his hand furtively covering his ear.

Left alone, Miss Stone settled at the teacher's desk, and had a look at the Detention Book. Only two names. She set her briefcase on the desk and opened it. Then her face went blank. It was a shock, slight but cutting, to realize that she had no correction to do. She liked to have work in hand when she was supervising a class. Then, if there was any disturbance, she could look up, severe and preoccupied. It gave her, she felt, an advantage.

She sat, red pen tapping the desk, surveying the little room, designed to depress further pupils who were already in disgrace; from beige walls to buff ceiling and brown floor. It was like sitting inside a parcel.

Outside, in the corridor were footsteps and voices, and the harsh, dangerous shouts of boys. They gradually died away. She wished with a fervour that surprised her, that somebody would come in.

The door opened, and a small, very dirty boy edged round it. He came and stood at the desk. He didn't look round the room or at Miss Stone. In fact, he gave the impression of not looking anywhere. His eyes were as blank as marbles.

'You're...let me see...Wayne. Is that right?'

'Aye.'

'Here's your work.'

She handed him a booklet and a sheet of paper, and he wandered off to a seat. He took some time to settle, pulling out a pen, then a pencil, and looking at them as if he had no idea what to do with them. Then he opened the booklet and began to write, slowly and carefully, squinting along the paper, his head lying along his arm.

The door burst open. The girl's face sagged in disappointment when she saw who was sitting at the teacher's desk.

'Aw!' she said. 'Mossy! I mean — sorry, Miss Stone. Where's Mr Downie?'

'I am taking Detention tonight,' said Miss Stone, in her colourless voice.

'But Mr Downie — he always takes it on Friday.'

'Looking forward to seeing him, were you? I see it was he who set you the Detention. Sit down, please.'

The girl slumped into a seat. She was about fifteen, pretty in a heavy, moist sort of way. The boy, Wayne, had not raised his head, but wrote steadily on. The girl's flushed face looked as if it might turn resentful and truculent. Miss Stone moved in, with the ease of long practice.

'What is your name? Ah, yes. Dawn. So, Dawn, you made yourself obnoxious in Mr Downie's class. He gave you Detention. And you were hoping to be alone at last with him, give or take one or two other numbskulls, was that it? Well, I am sorry to disappoint you. Mr Downie wished to see his fiancée this evening, and I agreed to take his place.'

'Fiancée!' The foolish mouth dropped, the eyes filled with tears. Dawn took the sheet of paper and sat, blinking and sniffing, making no effort to write.

'Go on,' said Miss Stone. 'A 500-word essay on Good Manners. And mind you write in paragraphs.'

Miss Stone sat, erect yet relaxed, hands clasped in front of her. She felt as she usually did, completely in control of the classroom situation. Then she was annoyed to realize that she was consciously achieving that control by an effort. 'For goodness' sake,' she said to herself, 'what's the matter with you? Another half hour and you can send them home. Then you can get away, as you planned.' She thought of her austerely comfortable little flat, and the glass of fine dry sherry she had promised herself.

It was no good. She was rattled. Why? It wasn't that silly girl, now sighing heavily as she considered the concept of Good Manners. No. It was the boy. There was something incongruous in the contrast between his dirty clothes and close cropped hair, his plastic jacket and heavy laced boots, and his earnest, scribe-like endeavours.

She went over to him. He laboured on, entirely absorbed in his work. He was copying from a booklet called *Spelling for Primary Schools*. There were lists of short, easy words...'able, ache, adder, affair'. His writing began large and ungainly, then trailed uncertainly upwards. Some of the words were wrongly spelt. She looked hard at the boy.

'What class are you in, Wayne?'

'2B Special, Miss.'

'How old are you?'

'Fourteen, Miss.'

She picked up his paper.

'Let me have a look. Good. That's very good, Wayne. "April" has a capital letter, though.'

'Oh, aye,' said Wayne, and resumed his task.

'How many of these are you supposed to do?'

'Fifty.'

'You've done far more than that. You can stop now. I'll put your work on the desk for Mr Price. He'll be pleased with you.'

She put out her hand for the piece of paper. To her surprise, he snatched it up and held it to him.

'I want tae dae mair words,' he said.

'No, no.' She was patient, in an impersonal way, because he was one of Mr Price's lot, in 2B Special. 'You've done enough.'

She stood, waiting for him to obey, but he sat, head bent, clutching the creased sheet of paper.

'Can I no dae mair words?'

Miss Stone was silent, puzzled.

'I like daen them!' His voice was suddenly loud and rough with defiance, and Dawn glanced up.

'I can dae it. You said I was daen it guid. I'm guid at it.'

He laid his paper on the desk, and smoothed it out.

Miss Stone opened her mouth to say something so sarcastic that it would sweep Wayne and his spelling list out of the Detention Room and out of her life for ever. But she did nothing of the kind. It was as if a veil had been wrenched aside, and a hard clear light shone on this dirty little boy. For the first time, in the long, cold years, she understood something...what he was saying...what he was telling her...why he was clinging to his pathetic task, so awkwardly done...that he was finding peace and satisfaction in the work. He was happy that he could cope. Everything in his school life was a fog of misunderstandings, a jungle of blundering errors, with no track to follow. Here, with his little book and his sheet of paper, he was like everybody else, successful and safe.

The moment of vision, with its pity, passed. To reassure herself, Miss Stone said, with a shaky laugh, 'You're not supposed to enjoy Detention, Wayne. You'll undermine the whole system.'

He sat silent, accepting yet again that he never knew what teachers meant.

Then he said, 'It's guid, this.' He gave a strong shiver. 'It's fine and warm, here. Quiet.'

He wiped his nose on his sleeve and began to write again.

Miss Stone became aware that Dawn was staring, mouth half open. She knew, by pupil instinct, that teacher had lost control of a situation.

Miss Stone decided to get rid of her.

'Have you finished, Dawn?' she said.

'Here, no Miss,' said Dawn, startled.

Miss Stone went to look over her shoulder.

'It'll do,' she said. 'You can finish it next Friday, when Mr Downie's here. I'll make a note of it in the Detention Book.'

'Oh, great.' Dawn got to her feet. Then her face darkened. 'I didna know that Mr Downie was engaged.'

'Did I say that? I must have been mistaken.'

'He's not engaged?'

'Not at all. He is, in fact, still looking for his ideal woman. Better luck next Friday, Dawn.'

She watched the girl making ready to go, clumsy and excited, like a puppy. She felt amused contempt for her, and an admiring awareness of her own malice and her...what had Mr Downie called it?...her sardonic humour.

At the door, the girl turned. She smiled radiantly, her face lit with love and joy.

'Thanks, Miss,' she said. 'You're great.'

Miss Stone felt that her moment of self-congratulation had been quite spoiled.

Left with Wayne, Miss Stone prowled round the Detention Room. While this wretched child was doing his Detention so conscientiously, *her* professional conscience directed that she should wait a little longer, even though all sound had died in the corridor and the Staff Room sherry would be long re-corked.

She glanced again at the Detention Sheet. Docherty...Docherty...a boy with the same shape of head and the same bristly fair hair.

'There was a Kevin Docherty,' she said.

Wayne jerked upright, and his pencil fell from his hand.

'That was my Dad, Miss,' he said.

'Miss What?' said Miss Stone, from force of habit.

'Miss Mossy.'

She shot a keen glance at him. Too stupid to be trying it on. He didn't look cocky; in fact, he was watching her almost as if he were frightened. Then his glance slid sideways.

'You tellt my Dad once he was as thick as a dry stane dyke.'

'Did I? I must have been annoyed with him. What's he doing these days?'

109

'Nothing,' said Wayne. He hesitated, blinking rapidly, then said, 'My Mam was in that class.'

'Oh? What was her name?'

'Janet.'

'What was her maiden name?'

Wayne looked blank.

'Her name before she married your Dad.' Wayne shook his head.

'Janet,' mused Miss Stone, '…Janet Stewart…That would be it, eh?'

'Dinna ken,' said Wayne.

'That would be in 1970. No, no, that would mean that Janet was only fifteen —' She stopped, embarrassed. There was no reaction from Wayne. He picked up his pencil ready to start that interminable copying.

'Have you any brothers or sisters, Wayne?'

'Twa brithers. Twa sisters.' He began to write again.

Miss Stone reflected on Wayne's parents. Although she always claimed she couldn't remember a thing about her ex-pupils, she had excellent recall. Janet Stewart, a wee fair girl, and Kevin Docherty….She remembered him all right. Big hands curled into fists, blue eyes congested with sullen hatred, while she stood over him, raking him with a sarcasm against which he had no defence except whispered obscenities. She had beaten him under, but only just. One of those boys whose violent fate has been decided and who are destined to decide the fate of others. Poor, silly wee Janet. Fifteen years old. She hadn't stood a chance. Miss Stone shrugged off Janet and her like.

'Your father was a bit of a hard man when he was at school. Did you know that?'

'Aye,' said Wayne. Then he whispered, 'He batters me. And my Mam.'

Miss Stone recoiled in alarm and distaste. Serve her right for showing interest. It was time the boy went home.

'Wayne, you've done two whole sides. Now, off you go.'

He didn't move. Sitting on the edge of her desk, her patient tone becoming ragged, Miss Stone said, 'Wayne, look, this has been my last day in this school. I don't teach here any more. I want to go home. And you must go home, too.' She took the booklet and paper from him and placed them on the desk. He watched her, picking at a filthy thumbnail. With a return of that eerie clarity of vision, she saw how white his face was, how hollowed his eyes.

'Get rid of him,' she thought. 'He's trouble. Don't ask him any more questions. You're on your way home…get him and yourself out of this room, out of this school…'

It was too late.

'I canna go hame. I'm feart. I want to stay here.'

Dreading the answer, Miss Stone had to ask.

'Why are you afraid?'

'It's my Dad.'

It took ten full minutes to get the story out of him. At the end of it, Miss Stone sat back, appalled.

The father had beaten the boy and his mother repeatedly. The other children were in care. Last night, things had been very bad. The woman and child had crouched in the kitchen, listening to Docherty in his drunken frenzy, raging above them. In the morning they had crept about, thinking themselves safe for a while. Then Wayne's father had flung open the kitchen door. The heavy, thudding blows on his mother had driven the boy beyond reason. He had picked up a heavy iron coal-shovel and struck at his father.

'It didn'a hurt him. It didn'a! Just his back. But he came at me and he said, "I'll kill you." An' I ran awa, an' he shouted after me, "When you come back the nicht, I'll batter you senseless! I'll kill you!" I was feart to go back…I cam tae the school.'

'And you haven't told anyone about this? A teacher?'

'Naw. Just you, Miss.'

Just her. Miss Stone sprang to the Detention Room door and looked the length of the corridor. No one in sight. No one to help. Everyone had done what she had wanted them to do, and gone home, leaving her with a boy, who had gone through a whole school day, half out of his mind with terror at what lay in wait for him at the hands of his father. 'Mam,' said Wayne behind her, 'My Mam…'

It was growing cold in the Detention Room. Outside, it was darkening, the wind was getting up, and rain was flung

against the windows. Distracted, she muttered, 'The police —
we could tell them —'

The boy started up in terror.

'No' the police!'

'But somebody must be told—'

Wayne's grimy hand clutched her arm, and even then she
shuddered with distaste, because no child had ever dared
touch her.

'No,' said Wayne. 'You come wi' me.'

'Come with you? Come with you where?'

'Hame. I'll no be feart if you come wi' me. Please. Please,
Miss Mossy.'

Miss Stone pulled her arm free, and went to sit at the
teacher's desk. She sat erect, as she had always done, contained and in command. But it was an outward show, a
parody. Within her raged tempests of fear and anger.

'It's not fair,' she moaned. 'I've retired...I've escaped...it's
nothing to do with me...Why should it be me? I've survived
everything because I never *never* got involved —'

The boy stood anxious, at her shoulder, and for a moment
she thought she must have cried aloud.

'Come on,' he said, and his face was old with dread.

Miss Stone was beaten at last. Defeated by the enemies she
had held at bay so long: pity, and stronger by far, compassion.

She got up, her movements random and slow.

'Where do you live, Wayne?'

'The Valley.'

'Of course,' said Miss Stone, drearily. 'Where else?' Hay Valley Gardens, that septic patch of decaying warrens.

They walked along the corridor. Once, Miss Stone stumbled
and nearly fell, for no reason.

'What will I find? Oh, what will I find?' Her mind threw up
ranging images of fear and disgust. Kevin Docherty, his great
granite hands, his face full of an old hatred...and a weeping,
beaten woman. Sordid, violent, ugly beyond belief...Among
the terrors of what was to come, there sounded, as from
another world, the voice of Mr Downie.

'I wish I had your sardonic sense of humour...you always
got your own way...'

Wayne looked up into her face.

'You're no' laughin' are you?'

'No. No. Not laughing.'

He put his hard, dirty little claw into her hand. She closed her fingers strongly over it, but it was the boy who drew *her* out into the rain and the wind and the dark.

The Teachers

they taught
that what you wrote in ink
carried more weight than what you wrote in pencil
and could not be rubbed out.
Punctuation was difficult. Wars
were bad but sometimes necessary
in the face of absolute evil as they knew only too well.
Miss Prentice wore her poppy the whole month
  of November.
Miss Mathieson hit the loud pedal
on the piano and made us sing
The Flowers of the Forest.
Miss Ferguson deplored the Chinese custom
of footbinding but extolled the ingenuity
of terracing the paddyfields.
Someone she'd once known had given her a kimono
  and a parasol.
Miss Prentice said the Empire had enlightened people
and been a two way thing.
The Dutch grew bulbs and were our allies in
wooden shoes.

We grew bulbs on the window sills
beside the frogspawn that quickened into wriggling
commas or stayed full stop.
Some people in our class were stupid, full stop.
The leather tawse was coiled around the sweetie tin
in her desk beside the box of coloured blackboard chalk
Miss Ferguson never used.
Miss Prentice wore utility smocks.
Miss Mathieson had a moustache.
If your four-needled knitting got no

further than the heel you couldn't turn
then she'd keep you at your helio sewing
till its wobbling cross-stitch was specked with rusty blood.

Spelling hard words was easy when you knew how.

To *Alexander Graham*

Lying asleep walking
Last night I met my father
Who seemed pleased to see me.
He wanted to speak. I saw
His mouth saying something
But the dream had no sound.

We were surrounded by
Laid-up paddle steamers
In The Old Quay in Greenock.
I smelt the tar and the ropes.

It seemed that I was standing
Beside the big iron cannon
The tugs used to tie up to
When I was a boy. I turned
To see Dad standing just
Across the causeway under
That one lamp they keep on.

He recognised me immediately.
I could see that. He was
The handsome, same age
With his good brows as when
He would take me on Sundays
Saying we'll go for a walk.

Dad, what am I doing here?
What is it I am doing now?
Are you proud of me?
Going away, I knew
You wanted to tell me something.

To Alexander Graham

You stopped and almost turned back
To say something. My father,
I try to be the best
In you you give me always.

Lying asleep turning
Round in the quay-lit dark
It was my father standing
As real as life. I smelt
The quay's tar and the ropes.

I think he wanted to speak.
But the dream had no sound.
I think I must have loved him.

VALERIE THORNTON

Only This

Jim has brought some dumpling, he said, there's a piece for you too.

Dumpling. The stuff you get in the butcher, shiny brown slices with liberal quantities of little black currants, hard and burnt tasting, sometimes with nasty little crunchy bits of stem or pip left in. Dumpling which the butcher gathers with hands bloody from mince and livers. He slaps the slices on the weighing machine's spattered surface and wraps them up with the raw sausages. Sometimes there's little red bits clinging to the cooked dumpling which you have to pick off. Dumpling with plastic round the outside, which you grill with the sausages and black pudding and which always burns hard and dry. The only dumpling now.

His mum made it for his father's birthday or something, he adds.

A vague memory stirs, but you let it rest.

Until you feel peckish and wander through to the kitchen. The dumpling's in the cupboard, wrapped in an opaque polythene bag, hidden from the cats. As soon as you open the bag, the smell seduces you, melts the years, dissolves the present. A smell unknown for what—you stop to count—ten, maybe twenty, years?

The mixed spices rise warm and friendly. The sultanas gleam quietly in the soft brown mixture of dried fruits and orange peel, flattened moon slices encircled by that wonderful thing—the skin of a clootie dumpling.

You loved them then. This one, Jim's Mum's one, is good. But it is only the key to the memory of heavenly ones.

When it was drawing near your birthday, you would ask Mrs MacDougall please to make you a dumpling. And she would give your mother a list of the things she needed because your mother never kept them in the house. Mixed spices—cinnamon, ginger and nutmeg, lots of sultanas, a bag of flour, a red, blue and yellow box of suet, and maybe other things. Little silver secrets, wrapped in greaseproof paper and hidden in the mixture—a thimble for an old maid, a horseshoe for luck, a ring for a wedding, a coin for wealth.

Mrs MacDougall would swing out the kitchen table with the red formica top into the middle of the room and put up the leaf so that she had room to work, with her sleeves rolled up, beating and mixing, raising a spicy cloud around her.

A clean linen dishcloth would be waiting nearby, the fine white kind with the blue or red borders. If you weren't already there, Mrs MacDougall would call you down when the bowl was ready to lick, when the mixture was tied up in the towel and simmering in a pan on the cooker. It would cook for hours, for all the morning, while Mrs MacDougall cleaned and ironed and washed and tidied and told you things. She knows everybody and tells you about them, about what a terrible time Mrs Thomson's husband gives her and nothing wrong with him but bad temper. She knows about Mrs Carn's shed which blew down and about who's died and who's had babies. She had six babies, six sons, six men, and she tells you about them and their wives and their children, her grandchildren. When she was a little girl she used to have long dark plaits, long enough for her to sit on. You laugh, because you can't imagine her as a little girl or with dark hair, this round, jolly person who has been this way since you came here. You were four and four days when you moved here and when you were four and five days Mrs MacDougall came to the door and offered to help you mother with the new baby, you little brother.

You would leave your room in a terrible mess. Your mother would rant at you to tidy it up so that Mrs MacDougall could clean it, but you knew that she would also tidy it for you if you didn't. Sometimes she'd tidy things away and you couldn't find them and you'd have to wait till the next day to ask her where she'd put them. But she was wonderful when you'd lost something. Have you seen this—my green scarf or my necklace with the elephant with the shiny pink stone or the hanky with my name embroidered on it? She always knows exactly where it is, and if she doesn't she will hunt until she finds it.

Some nights she comes and babysits, when your parents go off to a party or to a dance. Your mum will shimmer in before she leaves, beautiful as a princess in her down-to-the-ground shiny dresses. Midnight blue with diamanté, or ivory satin with pearls on the bodice, or your favourite with the pink and grey layers of net and the marcasite beading round the low neckline. You sometimes go and open the door of the down-stairs wardrobe in the spare bedroom and stand and gaze at the dresses hanging there. They seem unreal, magical, and the satin feels cold when you stroke it. They need your pretty mother's slim figure to bring them to life, her face perfect with powder and eyeshadow and bright lipstick, her long golden hair piled softly on top of her head that morning by the hair-dresser, her sapphire and diamond ring flashing on her finger, a cloud of soft perfume floating behind her. Sometimes the next morning there is a bouquet of flowers in the sink, exotic behind cellophane and satin bows, their raw stems, bruised and dark green, resting in an inch of water, waiting for Mrs MacDougall to arrange them in vases all over the house.

You have fun when Mrs MacDougall comes in the evening. She too has make-up on and jewellery. Pretty red beads and little gold earrings. She brings a handbag with her instead of her shopping bag and doesn't wear her overall. You know her and yet you don't know her like this. But soon you are laugh-ing—telling riddles and jokes and tongue twisters. She teaches you one about eenty teenty fickery fell, ell ell dominell, about Jack in the hazel tower. A hazel tower sounds wonderful. A tower is a magical thing and hazel is a creamy brown and hazelnuts are nice to eat and you know a girl called Hazel and she is pretty, not like you with your short straight hair and glasses.

Then she starts singing. She loves singing, and knows won-derful songs. They make you kind of sad inside too and you ask her to sing them and sing them until you can sing them too. You learn The Northern Lights of Old Aberdeen, which makes you sad, and when you ask her if that's where she comes from, if that's really where she longs to be, she laughs and tells you it's just the words of the song. You sing about the high road and the low road, and the bonnie bonnie banks of Loch Lomond. You sing about westering home with a light in the eye, back to my ain folk in Isla. But she doesn't come from Isla either, no more than you do. It's just the words of the song. You learn the descant for The Skye Boat Song at school and she loves it when you can sing together in harmony.

Then, long after your parents said you should be in bed, she begins to get you upstairs. She says get ready for bed and I'll come up and see you. You giggle and laugh a lot and plot tricks on her. One night, when she comes up the creaking stairs, you curl giggling under the covers. You have left a potty at the top of the stairs for her to trip over. But all she does is empty it for you and say nothing.

Then the bedroom games start. Wonderful games. You play The Minister's Cat, then I Spy, then you want to bounce on the bed. Then you play Throw the Sock Down the Back of the Bed. You have to get out of bed and sit at the bottom of it, with Mrs MacDougall in a wicker chair nearby. Any old sock will do, and all you have to do is hurl it at the dark head board until it falls down the back. You are all in stitches, because a sock is not very easy to throw. Sometimes it sticks and hangs over the back without falling down. When it does go down your little brother is always the one who has to go and get it. Mrs MacDougall takes a long time to bend down and reach in there, and you are scared of oose lurking in fluffy grey clouds on the brown linoleum. Your skinny little brother doesn't like going under the bed either, so he wriggles in and out as fast as he can.

Once, there is a sudden panic because you hear their car in the drive. Quick, into bed, she says, and puts the light out fast. She is downstairs and sitting by the fire on her own when the key turns in the door. You creep out of bed to hear what they say. No matter what you've got up to, she always says you've been as good as gold, and no bother at all. Her loyalty to you makes you all warm inside.

Sometimes you go to her house for tea. It's a real treat because she makes chips. Your mother won't let her do it in your house because she's afraid the kitchen will go on fire. You never have real chips at home. Chips and macaroni and cheese, and slices of buttered bread. And as a special treat you can eat in front of the television. Peyton Place is on, and you fall in love with Mia Farrow, you want to be her, pretty and elfin and with such lovely long hair. Sometimes you stay late enough for Coronation Street, with the lovely sad music. You like the lady with the sharp face and the hair net; even her name is sharp, Ena Sharples. She's so deliciously fierce, but it's safe because she's behind the screen and can't get at you.

When Mrs MacDougall goes on holiday to Blackpool she always brings you back a present, although your mother always tells her not to. You have a little wooden box, with varnished shells on the lid. You swing the lid sideways from one corner to open it. And a big lump of real crystal which is very heavy and clear and sits on the dressing table on a mat your gran crocheted. A green one with tiny roses worked into it. But your favourite is the little heart-shaped silver box with the red velvet lining. You keep your teeth in it. The ones the fairies didn't take away. One of them still has a speck of dried blood on it. It feels so strange when you put it into your mouth, not like it was yours at all.

Mrs MacDougall lives down by the river, but she doesn't like frogs. She's terrified of them and shudders uncontrollably when she tells you about the time she found a huge green one on the stone landing. She had to get a shovel and when it hopped on, she carried it all the way down to the back, so afraid it would jump off and land on her foot. You laugh and laugh at her, because you love frogs. You get tadpoles every spring but they mostly don't make it into frogs. One time you did have some little frogs but they hopped down the back of the washing machine which was bolted to the floor. You didn't tell anybody, especially not your mother, because you would have got a row and she would have got a man to take the machine out and look for frogs.

One summer you and a friend brought a lot of little baby frogs back from a pond. You had a sink in the garden and they could live there. But instead they crawled all over the rockery and were baked dry against the hot stones. Little flat frog shapes, grey and thin like burnt paper.

Another time, when you were at school, someone visited with a child, who went to play in the garden. The child knocked over your basin of tadpoles and Mrs MacDougall who hates them so much, was down on her hands and knees picking up the slippery little black things because she knew how you loved them.

She used to take your old bread for the hens. The hens were up on the wasteground beside her mother's house. Then when they built houses there, and the hens went away, she still took the bread. But she would bring potted hough which she had made herself for you. And always she would make a big pot of soup for the weekend, and wash and peel all the vegetables and leave them on top of the washing machine in a big bowl of water. She was so fast at peeling potatoes. You would stand by her elbow and watch her hand whizzing round a potato so fast you could hardly follow it. Then she would fish in the basin of cold water, thick with skin and earth, and bring out another dark potato to peel. You would talk about lucky potatoes, which she too remembered buying when she was little. Sweet cinnamon-dusted slices of delicious crumbly white stuff, with a gift concealed in it. She used to get silver threepennies when she was little, but now you get plastic soldiers or aeroplanes. You have to lick and suck and poke the sweet potato out of the tiny spaces on the gift.

She never eats sweets herself now; she used to ice cakes in a baker's shop and that put her off for life. But she likes pepper. She loves to see your face when she tells you she gets through a carton of pepper a week. And just to prove it, she always puts pepper on her cheese sandwich for elevenses.

One day when she was cleaning the stairs, there was a loud ripping noise—her whole knee had come through her stocking. You both laughed, shocked at the sight. Sometimes when she cleaned the stairs and you were at the bottom, you could see she wore peach coloured bloomers, but you didn't tell her because that was a private thing. She told you that she cleaned her knees with brillo pads and you laughed at the idea, but your mother was quite shocked.

When you went on holiday, she took the cats and the

goldfish. She loved to have the cats because they liked playing games with bits of bootlace and little balls of foil. She always had a lot of people around who could play with them. One time, one of the cats ate the fish and she was distraught when you got back. It didn't really matter, though, because fish are not the same as cats. You didn't even cry.

You left school and went away to study. She was still there, but you weren't. When you came back and saw her it was a little bit like strangers. She treated you not like you were a little girl, but like someone she had once known. You couldn't ask her for sweeties any more. She was a lot thinner and you saw her as someone between your mother and your grand-mother's ages. Then, your mother persuaded her to stop. To retire. You and your brother were both away from home and there was so little work now.

Once, you and your brother visited her. You stood on her doorstep, and you were all embarrassed. She was humble and respectful in a way you hated. She said you were a young lady now. You didn't want to be that kind of young lady. You wanted Peyton Place and macaroni and chips.

Then one year, the day after Christmas, your mother sees one of her sons pass the window. He comes to the door and tells you his mother has died. Suddenly. A heart attack. You are stunned. You don't cry. Not then.

It is snowing on the day of her funeral. You walk up the white road with your brother, slipping and giggling. Behind are your parents. Then you stop laughing when you get to the church and there are lots of cars, and a hearse, with its back door open, up in the air. There is no coffin. At the church door there are two wreaths of flowers on the wet tiles and inside the church there are lots and lots of people and a coffin on the floor in front of the pulpit. Suddenly it's all real and sad. You squash into the back row and wait for the last to arrive, her six sons and her husband, looking small, and bowed and old. As the service begins the snow starts to fall, thickly, past the gold and green and rose panes of glass. Mr MacDougall has asked for one of her favourite songs to be sung, and Cathie Jordan sings The Old Rugged Cross, and brings tears to your eyes and a huge lump to your throat. The minister says lovely things about her, and about how you should be glad to have known her and how there is a place for her in heaven, and you have to wipe your eyes and your nose. You try to think of something else, in this church decorated for Christmas with a tall green fir tree hung with shiny gifts and shimmering mists of angel hair and bright little strings of fairy light. The snow is swirling outside, beyond the pretty windows but you are inside and there are so many people mourning that crying seems to be the right thing to do, so you cry. Men are crying too. Your brother has forgotten to bring a hanky, so you give him a grotty one of yours but he doesn't seem to mind. You try to sing The Lord's My Shepherd, but you cannot sing. Your throat has closed up tight. You are trying hard not to sob.

You all stand as her six sons carry out her coffin. You are surprised at how small it is. They all look like her. You don't know whether this is happy or sad, but you cry anyway. Someone begins to sob loudly. The minister helps Mr MacDougall out of the church and everyone files out slowly, shaking hands with her sons, her sons in tears. The coffin is in the hearse, with flowers around it, with thick snow falling, covering the tracks and the footprints. You leave feeling desolate. It's the first time you've been together as a family for a long time. Your father takes you under his umbrella and offers you his arm. You have high-heeled boots and the snow is slippery, but you can't giggle about it any more. You hold his arm all the way home, feeling the warmth of your father by your side. It is the first time you have done this and it is comforting.

Although it was over already, it is really over now. You want only her to comfort you for this loss. But there is nothing.

Only this slice of dumpling, here, now, bringing her back to you.

But it hurts too much right now. You can always eat it later. It looks lovely, but not as perfect as her birthday dumplings for you. Nothing can ever match the perfection of the memory, the reality of then, in the sudden emptiness of now.

F*ollow On*

Old Wives' Tales

Reading and understanding

- The story is a monologue: who is the author talking to?
- Explain fully what you think the author means by saying it is 'the last day of my childhood'.
- What is making the 'Fat-Cat' so happy?
- What are 'grandmas' compared to and why?
- Explain why the writer's mother can't bear to watch war documentaries.
- What was the weather like on the days of the three births mentioned in the story? Why do you think the author mentions this each time?
- What do you think are 'all the other trappings of a statistic'?
- Why do you think the story is entitled 'Old Wives' Tales'?
- The opening sentence of the story is quite startling. It is often difficult to maintain the interest created by such an opening. How does the author try to do this and how successful is she? Discuss this in groups and then make notes on the conclusions you reach.

Talking

- 'The world out here's not a place to rush into. There's AIDS, drugs, and God knows what lurking out there, colds and rotten teeth, and acid rain and nuclear waste—all this in spite of the wonder of the modern world.' Either in groups or as a class, discuss whether, in your opinion, the World as it is now is a good or a bad place to be.

Writing

- The author suggests that books were of little use to her. Write about 100 words on some experience you have had which taught you something you could not have learned from a book.
- The story tells us a lot about the expectant mother. Describe the sort of mother you think she will be and explain why you think this.

126

The Doctor

Talking

● The doctor gives a variety of reasons as to why he carried out the abortion. What are they? Are any of the reasons justifiable? Can abortion ever be justifiable? Discuss these issues in groups or as a class.

Writing

● Explain the difference in tone and attitude between the evidence of the doctor and the monologue of the mother in 'Old Wives' Tales'. Pay particular attention to the differing ways they use language to express their point of view. Be sure to quote examples of both.
● Write a script of the conversation that may have taken place between the girl and the doctor before the doctor reached his decision to carry out the abortion.

Shadows Fall

Reading and understanding

● Why hadn't the mother gone to bed?
● Why can't she visualise what her own mother would have done in similar circumstances?
● How did the mother beat the curfew operating in her house when she was young?
● What word does the daughter use to describe her mother's friends and what does it suggest about her view of them?
● Explain as fully as you can why you think the mother describes her daughter as '...my mistake. Not my only one, but the one with the furthest reaching consequences.'
● Explain what the mother means by: 'I must steel myself for the worst.'
● What was it the daughter sometimes talked to her mother about and what was the mother thinking about so that she didn't often listen? What does this tell us about their relationship?
● What sort of future did the mother hope her daughter would have?
● We learn a great deal about the daughter's views on many aspects of life. Summarise these in as much detail as possible.
● Explain as fully as possible the last paragraph of the story.

Talking

Either as a class or in groups discuss the following:

● 'The problem with parents is that they forget that they too were

127

young.' How true do you think this is? Do parents interfere too much in the lives of their children? Does each generation face new problems that only that generation can solve?
• 'The sixteen or seventeen year-olds at school face more restrictions than their counterparts who have left school—their parents still view them as children.' Do you think this is the case? What differences are there between those still at school and those who have left?

Writing

• Put yourself in Marie's shoes. Write the story from her point of view, starting with her arrival at the house with Mike and ending with her reaction at finding her mother up and about at 6 a.m.
• A short story has to capture our interest quickly, hold that interest and provide us with a suitable ending. How well does 'Shadows Fall' satisfy these needs? Say how you think this story could be divided into a beginning, a middle and an end, and, with that in mind, write a paragraph on how effective you think each section is.

Poem for my Sister

Talking

Either as a class or in groups consider the following:
• Liz Lochhead wrote this poem when her sister was 12 and she was ten years older. She wanted to warn and protect her sister against what life had in store. It is almost as if she did not want her sister to grow up, and yet her sister wants to act as if she is older than she really is. Is it right that brothers and sisters should try to stop us experiencing life for ourselves? Should we be grateful for the benefit of their experience?
If you had a brother or sister at Primary School what sort of things would you warn them against? What things do you think they need protection from?
• In 'Poem For My Sister', what do we learn of the attitude of the poet towards her sister? After discussion make a list of the words and phrases in the poem that reveal the poet's attitude to us.

Flight

Talking

• Advertising, the media, and our peer-group all put pressure on us to want to be fashionable and one of the crowd. In groups or as a class, discuss how it is important to be fashionable and one of 'the crowd'.

Writing

● Read the poem again and pick out from the second verse the words and phrases which tell us most about the girl. With these in mind, write a short description of how the girl looked.

● Consider the feelings of Di's father, mother and sister. Write a paragraph about each of them trying to capture how they felt about Di's death.

● Imagine that Di, the girl in the poem, was a friend of yours. Write her a letter trying to persuade her not only that she is not fat, but also that people like her the way she is. Above all, try to convince her that she should give up dieting so drastically.

The Lost Boy

Reading and understanding

● On which night of the year does the story take place? Why is this important?

● Why was the boy so angry? Was Aunty Belle right to disillusion him?

● Why had the boy been punished at Halloween?

● How did the boy intend to rebel?

● How did the boy imagine Jock Scabra would be spending the evening?

● Describe in detail, and in your own words, what the boy saw when he looked through the window of Jock's cottage.

● Describe the face that the boy saw when he looked through the window.

● Why was the boy trembling when he went to bed?

● Read the last two paragraphs once more. Write down the words or phrases that the writer has used which emphasise the changes that have taken place since the previous night.

● Why do you think the 'one-eyed' cat has a prominent role in the story?

● In your own words try to describe the story-teller's attitude towards his Aunty Belle. Do you think the boy is right in feeling like this?

Talking

● 'Santa Claus, if he did exist, was a spirit that moved people's hearts to generosity and goodwill; no more or less.' Christmas has lost all religious meaning. We should accept that and treat it as it is— a holiday where large department stores and off-licence shops make a great deal of money and the population in general over-indulge themselves.

Would you agree with this point of view? Should we do more to ensure that the religious aspect of Christmas is seen to be observed? Should the churches get on with their celebrations and leave those who are not interested in this aspect of Christmas to get on with theirs? Discuss these points either in groups or as a class.

Writing

●   Write about an encounter you have had with an older person which you think helped you to grow up. Describe the encounter in your opening paragraph, and then explain the effect it has had on you.

●   You have been invited to make a three-minute Christmas appeal on television and radio on behalf of a charity. Write out the speech you would make, remembering that you have to persuade people to spend less on their family and friends and give money to help people less fortunate than themselves. Give your speech to the class and ask them how persuasive they think you were.

Away In Airdrie

Reading and understanding

●   What had awakened the boy?
●   Why did Danny's father and brother not go to the match?
●   What words or phrases let us know that Danny was keen to go?
●   What words let us know Archie is confident his team will win?
●   How do we know that Danny was reluctant to let Archie see him play football?
●Why did Danny get embarrassed when Archie introduced him to his friends?
●   What, for Danny, were the similarities between Broomfield Park and Hampden Park?
●   Why did Archie send Danny to a cafe?
●   Describe, in your own words, what Danny did from the time the cafe closed until he got on the train.
●   What thoughts do you think were going through Danny's head as he sat in the train going home?
●   An important punctuation feature is missing from the story. What is it, and why do you think the author has chosen to omit it?

Talking

● In view of what we, and they, knew about Archie, were the boy's parents right to let him go to Airdrie? Why do you think they did let him go? Should parents be more concerned about where their children are, and who they are with, when they are not with them? Discuss these points in groups or as a class.

Writing

● Write the script of the discussion which took place amongst the family when Archie eventually arrived home.
● Describe as fully as you can the character of Uncle Archie, as revealed by his actions and the attitude of the other characters towards him.
● Write an account of an event in your childhood which was special either because everything went right or because something went wrong.

Generations: Issues for Group Discussion

In all the following discussion points you should try to refer back to the various texts in this section and appoint one of the group to make notes on your conclusions.
● Why do older people fail really to understand the needs and motivations of the young?
● Can the young really learn about life from their elders, or must they learn from experience?
● What do you think are likely to be the most difficult things about being a parent?
● Do you think it is more difficult to be a young person now than it was in your parents' time? Why?

REALISATIONS

Flotsam and Salvage

Reading and understanding
● What is the girl doing on the beach?
● What are her feelings about what she is doing?
● How does the boy's attitude differ from that of the girl?
● What do you notice about the difference in how the two youngsters speak? What difference does this suggest about their backgrounds and characters?

- What was significant about their differing approaches to the job? Who was the more efficient and why?
- Why do you think that the two youngsters made a good partnership?
- Why is the discovery of the barrel of butter so significant?
- Describe, in your own words, the mother's reaction to her daughter bringing home the butter.
- Why do you think the girl had 'a sad sort of feeling' that Jeemie had left?
- How different is the girl at the end of the story from what she was at the beginning?

Talking

Either as a class or in groups consider the following:

- The girl assumes that Jeemie is of a certain character because of the way in which he talks and the way he dresses. How much can we really tell about people from their speech and dress? How important is it? Should we adopt different speech and different dress for different situations?
- What the girl comes to realise is the importance of material things. How dependent is our future contentment on money and material possessions?

Writing

- Imagine you are a reporter on a local paper. Write the story of the sinking of the ship (remembering to include eye-witness accounts) and the scene at the beach when the local people arrive to 'salvage' the cargo.

Morning Tide

Reading and understanding

- What was the boy doing on the beach?
- Why did the boy get so excited?
- Why did the boy suddenly decide to get on with the 'baiting'?
- Why did the boy feel he had been unjustly treated?
- Why did the boy experience pleasure when he was joined by Sandy Sutherland?
- Describe in your own words how Sandy treated Hugh. What were Hugh's feeling about this?
- Why did Hugh begin to feel differently towards his school friends?
- Why did the boys feel hostility towards Hugh?

- Describe Hugh's feelings about his bleeding nose.
- How did Hugh's character develop during the story, and what were the reasons for this?

Talking

Either as a class or in groups consider the following:
- Can anything be settled by fighting? Can physical violence ever be justified?
- Do you agree with Hugh that it is unmanly to cry? Why is it important to some men to have a 'macho' image? Why is it more acceptable for women and girls to cry?

Writing

- The extract from 'Morning Tide' describes an important incident in the life of Hugh. Describe the incident in your own words and explain why it alters Hugh's future.
- Describe an occasion in your life when you suddenly realised that you were no longer just a child but a young adult. Begin with a paragraph describing the occasion and then go on to explain the effect it had on you.

Fearless

Reading and understanding

- Through whose eyes are we seeing Fearless, and how does this affect our response to him?
- Explain in your own words other people's reactions to Fearless. How do you feel about these?
- Why do you think Fearless's glasses are described as 'terrible'?
- Why do you think people could not look Fearless in the eye?
- What reasons does the author have for saying that Fearless gave tramps a bad name?
- Why are CAPITAL LETTERS used in some places to indicate the spoken word while in other places *italics* are used?
- Why do you think that 'You were meant to think he was funny.'?
- Why does the author see some men as 'Romantics', and why does she make a sentence of that single word?
- Try to explain why the author felt both sick with fear and angry when Fearless approached.
- For what reasons do you think the girl's mother 'shook the living daylights' out of her?
- What do you think the author means by 'the hard, volatile maleness of the whole West Coast'? What part did men play in allowing

Fearless to continue acting like he did?
• What does the author realise as a result of her experience with Fearless?

Talking

• 'And I still see them smiling and ignoring it because they don't give a damn. They don't need to. It's not their battle.'
What areas of life do you see as holding advantages for men, and in what areas do women have the greatest advantage? Should those differences and advantages be allowed to continue? If you could change your gender, would you? What would you like most and least about being a girl/boy? Discuss these points in small groups or as a class.

Writing

• Drawing on what you know about Fearless, try to construct an account of how he came to be the way he is. Remember that what you say must fit in with the evidence we have.
• Was there a 'character' who frightened you when you were young? Write a description of the person, try to explain why you were frightened and say when, if ever, and for what reason the fear disappeared.
• 'It's a man's world, but women run it.' Write the speech you would make to a debating club or society either proposing or opposing the above motion.

Boys Among Rock Pools

Writing

• *Either* (a) Write a short story which involves hunting amongst rock pools where you discover something strange or unexpected;
 or (b) Write a description of a favourite beach or shoreline.
• Write a paragraph of *very detailed* description of someone who is concentrating hard on some task. As in the poem, try to focus on how their concentration affects body movement and expressions of the face and eyes.

Winter Bride

Talking

• Either as a class or in groups discuss the attitude of the fishermen who reported the accident and the response of his wife. Make a note of the evidence in the poem which supports your argument. Then go on and decide why you think the poet included the last line.

Writing

● Write a short story that involves an accident at sea.
● Put yourself in the position of the 'Winter Bride'. Write a letter in which she tells her mother about the incident, describing her feelings and what the incident will mean for her future.

Black Beasties that Bring Golden Reward

Reading and understanding

● From the information given in the opening paragraphs describe, in your own words, a lug worm.
● What is the best place to send for supplies of lug worms and why is this?
● Describe how one should *not* use a 'sprod' and why is this?
● Describe in your own words, the process of baiting and storing the fishing lines.
● What are *two* reasons why you must take great care when shooting the lines?
● Briefly describe the average working day of a line fisherman. Why did he work such long hours?
● Why was the author left on the pier one dark winter morning?
● Why didn't the author go straight to the doctor?
● Why did the doctor need his maid to assist him?
● What sort of fishing do you think the doctor did?

Talking

● 'Good old days? Ye must be jokin'!'
What things do you consider were better when you were younger? Are there things you miss doing because they now seem so childish? What aspects of your life do you feel are much better now? Do most things in life improve as you grow older?
Is there a historical era in which you would like to have lived? Why? Think about these ideas in groups or as a class.

Writing

● Using 'Black Beasties that Bring Golden Reward' as your source, write an article in the form of an interview between yourself and Peter Buchan.
● Peter Buchan was a fisherman for most of his life and this is reflected in the detailed description he gives of baiting and prepar-

ing fishing lines. Write a detailed description of one of the following activities which you have carried out:
  a) Repairing a puncture
  b) Preparing a meal
  c) Spring-cleaning a bedroom
  d) Clearing out the garage or the garden shed.
● After the removal of the fish-hook, the young seaman is given his first ever 'hurl in a car'. Write about an occasion when something unpleasant has happened to you, but some of the 'pain' is relieved by something pleasant happening as a result.

Realisations: Issues for group discussion

● Several of these extracts suggest that the sea brings hardship, but that this builds character. Is a certain amount of suffering and hardship good for us?
● Many of the characters in the stories come to realise that a living has to be earned. Is it possible to be unemployed and to lead a satisfactory life?
● The stories in this section suggest that the most important realisations come when we are young. Do you feel that your most important realisations are behind you?
● Having read and studied the realisations section, do you think that unpleasant experiences have a greater impact on us than pleasant experiences?

SEPARATIONS

The Bridge

Reading and understanding
● What are the other boys doing when they try to 'disqualify' the tiddler? Why are they doing this?
● How has the hero's relationship with the other boys changed after he shows them the tiddler?
● What do you think is meant by 'spanning' the bridge? You may find drawing a picture that shows how the bridge fits into the landscape will help you to visualise the scene.
● Do you think it was fair to ask the boy to span the bridge? Give a reason for your answer.
● Explain, in your own words, what is going on in the boy's mind

as he goes across the river.
- Why does the fish that he caught no longer seem to matter?
- Why do the other boys run away, and does it really matter to the hero?
- The story describes a simple incident, but why is it important to the boy?

Talking

- When we are young we are often 'dared' to do things to prove to others that we are not 'chicken'. Prepare a three-minute talk on a time when you were dared to do something. Describe how it came about, what you did and how you felt during and after the event.
- As we grow up do we become more tolerant and considerate to each other, or do we merely develop more subtle and sophisticated forms of cruelty? Discuss this question in groups or as a class.

Writing

- Describe some situation when you were cruel or unkind to someone else and try to account for your behaviour.
- Spanning the bridge was obviously considered to be quite an achievement by the boys—almost a test of manhood.
   Either (a) write a story which involves a similar type of 'test';
   or (b) if you have ever had to prove yourself in a similar way, explain the circumstances, describe what you had to do and explain the outcome.

A Scottish Lion Rampant

Reading and understanding

- Why does John mention the fact that his father served in the British Army?
- What do you think is meant by 'double shifting'? What does it tell us about the two brothers?
- Why was going to school an unpleasant experience for the boy? Whose fault do you think it was?
- Was his father's advice to him, not to fight, good advice?
- Why do things improve for him at school and what effect does this have on relationships within his own family?
- How do the Scottish boys differ from the immigrant boys at secondary school, and why do you think this is so?
- What is your opinion of the Sikh initiation ceremonies?
- Describe, in your own words, how the television engineer treats John when he goes for an interview.

- Do you think John is right to interpret his workmates' treatment of him as involving 'nothing nasty' and 'natural'?
- To what extent do you agree with John's interpretation of Scottish life?

Talking

- As a class or in groups think about whether there ought to be stricter control of immigration into this country.

Are there advantages to living in a multi-cultural society? If people come to live in this country should they give up their own language and customs and adopt ours?

Writing

- Read John's description of his interview with the television engineer again. Now write a script of the dialogue that may have taken place.
- 'Britain is a country in decline: in population, in industry, in morale.' Do you consider this to be true? In at least 100 words, explain your opinion and try to think of examples that support your point of view.
- John Singh tells of some of the difficulties *he* found in trying to get a job. What are the qualities which you personally have to offer an employer, and what are the things that will work against you when you go to seek a job? Write one paragraph on your strengths and another on your weaknesses. *Be honest*!

The First Men on Mercury

Reading and Understanding

- What prevents communication between the earthmen and the men from Mercury?
- What has happened to each of the groups by the end of the poem?

Talking

- Either as a class or in groups consider whether space exploration is a waste of time and resources. What benefits are we likely to receive because of the space programme? In a world where so many people seem to suffer famine and other natural disasters, is there any way we can justify spending billions of pounds each year on space research?

Wer Da?

Reading and understanding

- What does the name 'Herman' tell us about what being German meant to these pupils?
- What other details in the first verse confirm this?
- How effective was 'German Geordie' as a teacher?
- How does the discovery that 'German Geordie' was an Austrian Jew affect the poet?
- Why do you think the poet added the last two lines of the poem?

Writing

- The poem has a great deal of military imagery and words associated with war. Make a list of all words with military associations.
- 'German Geordie's' fault, in the eyes of the boys, was simply that he was different. Write down the thoughts that you think might have been going through his head as he once more prepared to face his class one morning.

The Jam Jar

Reading and understanding

- Why did the narrator like catching bumble bees or wasps in a jam jar?
- How do the wasps behave when released?
- What had happened in the City of Aberdeen in the summer of 1964?
- In what ways does the author come to be like a bee trapped in a jar?
- Why do 'waspish' women behave as they do?
- How does the author feel about the way people in hospital behave to each other?
- Why does she break the jam jar at the end of the story?

Talking

Either as a class or in groups consider the following:
- Re-read the first two paragraphs of the story. Do they provide a good opening for the story or could they be omitted?
- The story suggests that 'human nature never changes, always, always, there will be victors and persecutors.' Do you think that this rather bleak description of human nature is true? Think of instances

of human behaviour that support this statement, but also instances that contradict it.

What is it that brings out the worst in human nature? Could we change society in some way to bring out the best?

Writing

● Words associated with insects are used throughout the passage. Working in groups go through the story identifying these phrases or words and write them down. Re-group as a class and compare your findings.
● Imagine you have been trapped or confined in an enclosed space. Try to describe your feelings in this situation.
● We have, all of us, been ill at some time. Describe a time when you have been ill, paying particular attention to what you felt (towards yourself as well as others!) during this period.

Greensleeves

Reading and understanding

● What significance does the tune which the ice-cream van plays have for the old woman?
● What, as far as the old woman was concerned, were the advantages and the disadvantages of living in the tower block?
● What kind of relationship does the old woman have with her neighbours?
● What were the good moments and what were the bad moments of using the lifts on a weekend and why does the woman like weekends best?
● What do you think of the comments of the women on the bus?
● What do you think of the way the woman has been re-housed and what are the main differences between her former home and her home in the tower block?
● What were the advantages and disadvantages of *not* having a coal fire?
● Why do you think the woman took such good care of her plants, even the weeds, and what does that tell us about the woman?
● Do you think that the neighbour who comes to her door is really all that different from her? How is he going to spend the evening and why?
● What do you think is the significance of the television money running out just as the minister had finished his first sentence and what does the woman's reaction to this tell us about her?
● Why does the author give us details about what is on television?
● What is the lock salesman trying to do to the woman?

- What do you think is important about the woman watching the film, *Grand Hotel*?
- Why do you think the author called the story 'Greensleeves'?

Talking

- In groups or as a class, discuss who you think is to blame for this old woman's predicament. What could be done to improve life for her? Would she be better off in a home?

What is the evidence in the story that highlights the loneliness of the woman? How important do *you* think it is to have a lot of friends?

Do you think elderly people like company or do they prefer their own company most of the time?

How important do you think happy memories of the past are? Do older people spend too much time thinking about the past and not enough considering the future?

Writing

- Write a story that takes place in the next few days and involves the woman with either a neighbour or the lock salesman.
- 'Greensleeves' is the tune that helps revive memories for the old woman. Is there a song or piece of music that brings back memories (happy or sad) for you? If so, say what the song or piece of music is and describe in detail the circumstances and what memories it brings back.

Separations: Issues for group discussion

- Is there something wrong in being an outsider or individualist? What are the advantages of conforming and being one of the crowd?
- We all have stereotyped notions of what foreigners are like. Consider where these notions come from and whether they do any harm. For example, what are French people like? What do you feel about Chinese people? Are all Scots mean?
- Many of the extracts in this section have highlighted cruelty, indifference, prejudice and lack of human caring. Think about what has been in the news recently. What evidence has there been of these things and what evidence has there been that there is a better side to human nature?
- Soon you will be moving from childhood into the adult world. What aspects of the adult world do you think you will be able to cope with, and what aspects fill you with apprehension?

CONVERSIONS

The Time my Friend Mabel became an Anchorite

Reading and understanding

- What confusion arose in Phil's mind when his wife decided to become an anchorite?
- What do you think had inspired Mabel to become an anchorite?
- Why did the author think that sending for the doctor would do no good?
- Try to explain, in your own words, how the Americans behaved when they came to see Mabel.
- What impression is the author trying to create in her interview on 'News at Ten'?
- What do you think of Phil's general reaction to Mabel's conversion?
- List the evidence in the passage that lets us know that all different sorts of people were interested in Mabel's withdrawal from the world.
- Mabel has gone to Novaya Zemlya, which is in Russia. What sort of place do you think this is and why do you think she has gone there?
- What, in the end, were the advantages brought about by Mabel's disappearance to Phil and the author?
- What does the way in which the author tells the story tell us about her and other people's attitude to Mabel's conversion?

Talking

- 'I've had enough of people. I'm going to spend the rest of my life in total solitude and silence contemplating the infinite.' In groups or as a class, think about whether any of us has the right to withdraw totally from society, or whether we should be doing more to improve the world by getting involved. Has anyone the right to total privacy? Should everyone share the burdens of the world equally?

Writing

- Write about the same events as though you were *either* Mabel *or* Phil. Don't forget to show their view of the character who is the author in the original version.
- Quite a lot of humour in the passage is achieved by contrasts. Make a list of as many of these humorous contrasts as you can.
- Few, if any of us, want to cut ourselves off completely from our friends and the outside world, but there are times when we all like

142

to get a bit of peace and quiet, be on our own and relax or quietly work through our problems. Try to explain on what occasions you do this, where you go and why you try to find a corner of peace and solitude.

Snowballs

Reading and understanding

- Why did Mr Macrae want the boys to be on their best behaviour?
- What evidence is there that Kenneth soon got used to the playground activities?
- Describe, in your own words, the thoughts that go through Torquil's head as he waits to be punished.
- Account for and describe Mr Macrae's mood change from the time Torquil returns to the classroom up until Torquil leaves.
- What part of his story did Torquil have difficulty telling his father about? Why do you think this was?
- Why did Torquil decide to emigrate to Canada?
- What thoughts do you think went through Torquil's head as the ship pulled away from the shore?
- We are given several details of Torquil's early life in Canada. Which do you think was the most unpleasant feature and why?
- Why does the author include reference to the argument with Smith at the ceilidh?
- Why do you think Torquil never went to church? Was his attitude justifiable?

Talking

Either as a class or in groups consider the following:
- Torquil obviously resented the treatment he had received in school and it stayed with him. Do you think he was right to adopt this attitude? Do you think he should have been more forgiving? How long can one go on holding a grudge?
- 'One problem with teachers is that all too often they never really listen to the pupils' point-of-view.' Do you agree with this statement? Why do you think this is? Do you think that if teachers knew what pupils thought about, say, the running of school, they would change any of the rules or alter their behaviour? Is it just that teachers don't listen to pupils, or is it also that *adults* aren't really interested in what young people think?

Writing

- Torquil felt that he was unjustly punished. Write about a case of injustice where someone has been punished for something they did not do. This can be based on an actual event or you can write a short story with this as the theme.

- Write a letter to your local newspaper complaining about some aspect of life where you think teenagers are unfairly treated. For instance they may be prevented from using shops, cafes or clubs because of the way they dress or just because they act in a way that adults find noisy. Explain your complaint and say how you would like to see it put right.
- 'he could hear the words of that poem.' — 'A Wet Sheet and a Flowing Sea' by Allan Cunningham is the poem the boy will always remember. Is there a poem you have studied in school that you will remember? Give a brief description of the poem and then explain what it is about the poem that will make you remember it.

Four of the Belt

Reading and understanding

- Who is talking and who is he talking to?
- What do you understand by the phrase 'ritual physical humiliation'?
- Why do you think the speaker says:

>   and if you cry, boy, you will prove
>   what I suspect—you are not a man.

- Explain the phrase 'as they say '.
- What do you think are the implications of 'But I show you I am a man, by doing this, to you.' What does this tell you about the speaker?
- What does the speaker think Jenkins will learn from being given four of the belt?
- How does the writer think Jenkins will react in future years?
- Explain the last line of the poem.

Talking

- Either as a class or in groups, discuss whether punishment needs to be physical. What punishment or restrictions should be applied to people who offend against school rules and classroom discipline? Do you feel they would be effective?

How can we ensure that a minority of pupils don't make life unpleasant and disrupt the education of the majority?

Writing

• Does the poem 'Four of the Belt' shock or amuse you? Try to explain why you have this reaction.
• Imagine that you have been stopped by someone in authority and accused of doing something you didn't do. Write a script of the dialogue between yourself and your accuser, giving details of your alleged 'crime' and how you manage to prove your innocence.

Mossy

Reading and understanding

• Consider the pupils' actions from the time Miss Stone lined them up until the time they left the school premises. What does this tell you about Miss Stone as a teacher?
• What reasons does Miss Stone give for wishing to take Mr Downie's detention? Explain why 'Miss Stone waited, looking down her nose.'
• Why did Miss Stone like to have work at hand when she was supervising a class?
• Dawn is disappointed to see Miss Stone taking detention. What were her feelings and how did they change from the moment she came into the classroom until Miss Stone told her to write a 500-word essay on Good Manners?
• Why was Miss Stone rattled and what was the surprise outcome of telling Wayne that he could stop? What did Miss Stone mean by saying, 'You're not supposed to enjoy Detention, Wayne. You'll undermine the whole system'?
• Why did Dawn feel it was 'great' having to do detention the following week?
• Explain, in your own words, what Kevin Docherty had been like when he was at school.
• Explain clearly what had happened in Wayne's house the previous night.
• What does the author mean by saying that Miss Stone had held compassion 'at bay'?
• Why do you think Wayne turned to Miss Stone for help rather than one of the other teachers? In what way had Miss Stone been defeated at last? Why do you think the author called the teacher Miss Stone?

Talking

Either as a class or in groups consider the following:
• Part of the enjoyment of 'Mossy' lies in the author's skill in pre-

senting us with a memorable character in relatively few words. Using the story, explain what sort of person you think she is. Do you think she is a different person at the end of the story?

● Wayne was determined not to involve the police in his troubles. Do you think he was right? Why do you think some people avoid contact with the police in this way? Where do you think Wayne and those like him should seek help? What steps do you think society at large should take to help protect people like Wayne and his mother?

Writing

● Quite often we find our views about a person change once we get to know them better—sometimes we ourselves are changed by our contacts with others. Write a short story that involves an encounter that seriously changes the views of one of those involved.

● Mossy did not want to be responsible for Wayne. Write about a time, perhaps when you have been forced to look after a younger member of the family, when you had to take on some responsibility in spite of the fact that you did not want to. Describe what you did, and explain any problems that may have arisen and how you felt about it afterwards.

The Teachers

Talking

● Either as a class or in groups discuss whether schools really teach pupils what they need to know. What subjects need to be taught now? Are there any subjects that need to be brought up to date or got rid of altogether?

Writing

● The poet is recalling things she learned in school and the things that have stayed in her mind, the things that help to shape our lives for good or bad. What things that you have learned in school do you think you will remember in later years and why? Have you learned anything which you think will help you in the future? What will help you most?

● Write a speech that is in favour *or* one that is against the motion: 'Wars are bad but sometimes necessary.'

To Alexander Graham

Talking

● Either as a class or in groups, discuss what expectations parents have of their children. Is it a good or bad thing that your parents expect you to achieve certain standards?

Your parents provide for you; have they a right to expect anything in return? Do you take your parents for granted, forgetting that their concern is only natural? Do you think you expect enough of yourself?

Writing

• The poet has only come to realise his true feelings towards his father late on in his life. Quite often we hide our feelings or are unaware of what our true feelings towards someone or something are. Write about any situation where you have not realised exactly what your feelings were towards someone or something, until too late.

• The poet recalls places he visited with his father. Write a description of somewhere you remember visiting when you were younger, emphasising the features that stand out most vividly in your memory.

Only This

Reading and Understanding

• Who is the writer talking to in the story and how do we know this?

• What are the things she dislikes most about the dumplings you get from the butchers?

• Explain what you think the writer means by: 'As soon as you open the bag, the smell seduces you, melts the years, dissolves the present.'

• What can you tell from the fact that the mother never kept the ingredients for dumplings at her house?

• What were the good things about being around Mrs MacDougall when she was making dumplings?

• What are the significant differences between the way her mother dresses on party nights and the way Mrs MacDougall dresses? How does this reflect the way the girl feels about them both?

• Why do you think the children particularly enjoyed the things Mrs MacDougall did with them when their parents were out?

• Explain as fully as you can the 'loyalty' of Mrs MacDougall and why was it important to the writer.

• Why do you think the writer recounts the stories of the frogs and tadpoles?

• 'You left school and went away to study. She was still there, but you weren't.' Explain as fully as you can what you think is meant by this statement.

• 'She was humble and respectful in a way you hated.' Try to

147

explain why Mrs MacDougall was like this and why the writer hated it.
- How is the scene outside the church contrasted with what is going on inside?
- What do you think is meant by: 'Although it was over already, it is really over now.'?
- Why do you think the story is called 'Only This'?

Talking

Either as a class or in groups discuss the following
- 'The real appeal of Mrs MacDougall stems from the fact that she belongs to a different social class from the author.' Do you think this is true? What are the significant details in the text which tell us that Mrs MacDougall has a different life-style and different attitudes from the author's family?
- 'Older people are always talking about the past. The past is gone, you cannot change it. The future is what is important.' Do you agree that the past is only worth forgetting and that it should play only a minor role in our lives? Do you think that thinking too much about the past is damaging to the future? How much does the past influence the future?

Writing

- Coming across something you have not encountered for a long time (like a piece of dumpling!) can often bring back memories of people or places. Describe a situation where something has triggered off memories for you in this way.
- Did you have a favourite food when you were younger? Say what it was and try to describe your feelings as you waited to get it, what you felt while eating it, and what you felt like after you had eaten it.
- People whom we are involved with when we are young often leave a lasting impression on us. Describe your memories of someone who was very important to you when you were younger.

Conversions: Issues for Group Discussion

- Most of the characters in this section are involved in unpleasant experiences which stay with them for the rest of their lives. Do you think we learn more from the unpleasant things that happen to us, and why do you think this is?
- What do you think are the advantages and the disadvantages of being a school teacher?
- Discipline in schools is less severe than it was. Has this produced a better atmosphere for learning? Do pupils prefer teachers who

insist on strict discipline, like Mossy?
● The stories in this section seem to suggest that changes in people's thinking come at moments of sudden conversion. Do you believe this is true? Has there been an occasion when this has happened to you?

General Follow-on Activities

● Many short stories are dramatised as television plays. Which of the short stories in this collection do you think would make a good television drama? In small groups make a list of the characteristics of a story which make it appropriate for television. Now, as individuals, chose one story and re-read it closely, then try to re-write it as a play suitable for television.
● The author of a short story often uses suspense to build up a reader's curiosity and then holds back from satisfying that curiosity until the end of the story. Write about how an author has done this in any of the short stories in this collection, and say whether or not you think they have done so convincingly and to good effect.
● All the stories in this book are by Scottish authors, or authors who now live in Scotland. Of the stories you have read which is the most 'Scottish', and why? Is it only a matter of language or are there other factors that go to make up 'Scottishness'? Which of the stories is least Scottish? Explain why you think so.
● In 'An Approach to Writing', Sheena Blackhall refers to herself as a 'visual' writer. Imagery is of great importance to her. Is there another writer in the collection whom you would describe as a 'visual' writer?
Briefly describe the story you have chosen. Then, in some detail and with close reference to the story, explain why you think that the author is a 'visual' writer.
● What do you think are the ingredients of a really worthwhile poem? Consider again any of the poems in this collection and explain clearly why you think it has some of the ingredients that you think are important and why you feel that it is worth reading.
● You have read and considered many poems—now, write your own on 'Relationships'. Remember to think about such things as structure, setting, rhyme, imagery—in fact all those things you have been thinking about in other poets' work.
● Write a short story which focuses on childhood. The 'hero' is to be a young boy or girl, once very active, who, as a result of an accident, is now handicapped and is always being left behind. Write in the first person (ie 'I...') and try to capture their feelings as sensitively as you can.

Group Investigation and Report

The short stories and poems in this collection have been about people of all ages, and in all sorts of situations.

Each member of your group should interview someone from a different age-group (it is important to cover as wide an age span as possible).

It is important that before you begin you should discuss as a group the things you would like to find out about, remembering not to be too personal. Find out about people's dislikes: their attitude to being young or being elderly; retirement; nursery schools; playing games (remember, adults play games too!).

What effect does the arrival of a baby have on a family? What was it like in school forty or fifty years ago? What was entertainment like then? What foods were eaten? What do young people think it is like being old? There is an endless list of questions one can ask, but do not ask too many. Try to tape record your interviews.

Having recorded and assessed your material produce a group report on your findings, and report to the class.

Finally, the class as a whole should produce a report and, perhaps, come to some conclusions about how people change as they grow older and how much they remain the same.

(N.B. Don't try to keep people to too rigid a line if you are tape-recording them. People will find it easier to talk if the atmosphere is informal. One can gain a lot of insight into people's thoughts just by listening to them talk.)

Unwin Hyman English Series

Series editor: Roy Blatchford
Advisers: Jane Leggett and Gervase Phinn

Unwin Hyman Short Stories

Openings edited by Roy Blatchford
Round Two edited by Roy Blatchford
School's OK edited by Josie Karavasil and Roy Blatchford
Stepping Out edited by Jane Leggett
That'll Be The Day edited by Roy Blatchford
Sweet and Sour edited by Gervase Phinn
It's Now or Never edited by Jane Leggett and Roy Blatchford
Pigs is Pigs edited by Trevor Millum
Dreams and Resolutions edited by Roy Blatchford
Shorties edited by Roy Blatchford
First Class edited by Michael Bennett
Snakes and Ladders edited by Hamish Robertson
Crying For Happiness edited by Jane Leggett
Funnybones edited by Trevor Millum

Unwin Hyman Collections

Free As I Know edited by Beverley Naidoo
Solid Ground edited by Jane Leggett and Sue Libovitch
In Our Image edited by Andrew Goodwyn
Northern Lights edited by Leslie Wheeler and Douglas Young

Unwin Hyman Plays

Stage Write edited by Gervase Phinn
Right on Cue edited by Gervase Phinn
Scriptz edited by Ian Lumsden

*F*urther Reading

A Nippick o' Nor' East Tales – Sheena Blackhall (Keith Murray)

The Sun's Net – George Mackay Brown (The Hogarth Press)

Hawkfall – George Mackay Brown (The Hogarth Press)

Andrina – George Mackay Brown (The Hogarth Press)

Fit Like Skipper – Peter Buchan (Aberdeen Journals)

Fisher Blue – Peter Buchan (Aberdeen Journals)

Sun Circle – Neil M. Gunn (Faber & Faber)

Butcher's Broom – Neil M. Gunn (Faber & Faber)

Morning Tide – Neil M. Gunn (Faber & Faber)

Mr Trill in Hades – Iain Crichton Smith (Gollancz)

Murdo and other stories – Iain Crichton Smith (Gollancz)

Consider the Lilies – Iain Crichton Smith (Gollancz)

The MacMillan Companion to Scottish Literature – ed. Trevor Royle

A Companion to Scottish Culture – ed. David Daiches (Edward Arnold)

The Concise Scots Dictionary – ed. Mairi Robinson (Aberdeen University Press)

The following publications appear on a fairly regular basis and are well worth seeking out.

Original Prints – *New writing from Scottish women* – published by Polygon, Edinburgh.

Scottish Short Stories – published annually since 1973 by Collins Ltd.

New Writing Scotland – published annually by the Association of Scottish Literary Studies, University of Aberdeen.

Acknowledgements

The editors and publisher are very grateful to the following for permission to reproduce the material in this book:

'Old Wives' Tales' © Wilma Murray, first published in *Original Prints* Vol. 2, Edinburgh University Press, 1987.

'Stobhill – The Doctor' and 'The First Men on Mercury' © Edwin Morgan, taken from *'Poems of Thirty Years'* by Edwin Morgan, published by Carcanet Press Ltd.

'Shadows Fall' © Pat Gerber, first published in *New Writing Scotland* No. 3.

'Poem for my Sister' © Liz Lochhead, from *Memo for Spring*, published by Gordon Wright Reprographia.

'Flight' © Ken Morrice, taken from *When Truth is Known*, published by Aberdeen University Press, 1986.

'The Lost Boy' © George Mackay Brown, taken from *Andrina and Other Stories*, published by Jonathan Cape Ltd.

'Away in Airdrie' © James Kelman, taken from *Not While the Giro and Other Stories*, published by Polygon, 1983.

'Flotsam and Salvage' ©Enid Gauldie, taken from *Scottish Short Stories 1984*, published by Collins Ltd.

'Morning Tide' © Neil Gunn, from *'Morning Tide'*, published by Souvenir Press Ltd.

'Fearless' © Janice Galloway, first published by *New Writing Scotland* No. 6. 'Boys Among Rock Pools' © George Bruce, taken from *Collected Poems of George Bruce* published by Edinburgh University Press, 1971.

'Winter Bride' © George Mackay Brown, first published in *Fishermen with Ploughs* by The Hogarth Press.

'Black Beasties that Bring Golden Reward' © Peter Buchan, from *Fit Like Skipper* by Peter Buchan, reproduced courtesy of Aberdeen Journals Ltd., (Evening Express), 1985.

'The Bridge' © Jessie Kesson, taken from *Where the Apple Ripens*, first published by The Hogarth Press.

'A Scottish Lion Rampant' © John Singh, first published in *From Where I Stand* edited by Desmond Mason, published by Edward Arnold.